Studies in Fuzziness and Soft Computing 288

For further volumes:
http://www.springer.com/series/2941

Studies in Fuzziness and Soft Computing 288

For further volumes:
http://www.springer.com/series/2941

Rafał Scherer

Multiple Fuzzy Classification Systems

Author

Dr. Rafał Scherer
Department of Computer Engineering
Czestochowa University of Technology
Poland

ISSN 1434-9922 e-ISSN 1860-0808
ISBN 978-3-642-43657-4 ISBN 978-3-642-30604-4 (eBook)
DOI 10.1007/978-3-642-30604-4
Springer Heidelberg New York Dordrecht London

Softcover reprint of the hardcover 1st edition 2012

Printed on acid-free paper

Springer is part of Springer Science+Business Media (www.springer.com)

Preface

Exploratory data analysis is a vital set of methods used in various engineering, scientific and business applications. Fuzzy classifiers are important tools in this growing field. They use fuzzy rules and do not require assumptions common to statistical classification.

Rough set theory is useful when data sets are incomplete. It defines a formal approximation of crisp sets by providing the lower and the upper approximation of the original set. Systems based on rough sets have natural ability to work on such data and we do not have to preprocess incomplete vectors before classification.

To achieve better performance than existing machine learning systems, we combine them in ensembles. Such ensembles consists of a finite set of learning models, usually weak learners. In the book two popular methods are applied – boosting and negative correlation learning. Both of them help to achieve better accuracy and are not prone to over-fitting.

This book combines three aforementioned fields – fuzzy systems, rough sets and ensembling techniques. As the trained ensemble should represent a single hypothesis, a lot of attention is placed on possibility to combine fuzzy rules from fuzzy systems being members of classification ensemble. We build ensembles of various neuro-fuzzy systems with certain modifications to let the fuzzy rule bases to be activated at the same level. Furthermore, a lot of emphasis is placed on ensembles that can work on incomplete data, thanks to rough set theory. Presented ensembles give information how the number of missing features influences classification accuracy.

First of all I would express my sincere thanks to my friends Dr Marcin Korytkowski and Prof. Robert Nowicki who provided me help in research and to Prof. Leszek Rutkowski, who introduced me to scientific work and supported me in a friendly manner. I am also grateful to the Department of Computer Engineering at Częstochowa University of Technology for providing a scholarly environment for both teaching and research. Finally, I am truly grateful to my wife Magda for her love and patience and to my mother for raising me in the way that she did.

Częstochowa, Poland

Rafał Scherer
April 2012

Acknowledgements

This book was partly supported by the Foundation for Polish Science – TEAM project 2010-2014 (TEAM/2009-4/9) and by the Polish Ministry of Science and Higher Education (Habilitation Project 2007-2010 Nr N N516 1155 33).

Contents

Chapter 1
Introduction

Classifying objects described by a set of their numerical features is one of the basic tasks of pattern recognition and data mining. It is applied in many domains such as medicine, economics, fraud or fault detection, etc. Designing better classifiers is a subject of sustained research and through the years many classification methods were developed [2, 3, 4, 7, 11, 19, 13, 29, 30, 31]. The most popular ones are those based on i.a. neural networks, nearest-neighbour, decision trees and support vector machines. There does not exist one best classification method. We can choose only one, best classifier for a given task using laborious trial and error method but in this way we miss collective knowledge of discarded classifiers. To address this problem many methods for automated aggregation of classifiers trained for the same task have been developed [12, 14, 24]. These methods demonstrate that the classification accuracy nearly always improves after combining different classification methods or classifiers trained with different datasets. Nowadays combining single classifiers into larger ensembles is an established method for improving the accuracy. Of course, the obvious improvement is bound up with increased storage space (memory) and computational burden. However this trade-off is easy to accept with modern computers.

The book applies fuzzy, neuro-fuzzy and neuro-rough-fuzzy ensembles in the classification problem which is of utmost importance in virtually all fields, notably in decision making, pattern recognition, etc. In the book two challenging problems will be solved. It will be shown how to join fuzzy rules from all subsystems creating an ensemble and how to design an ensemble of fuzzy subsystems in the case of missing data. Throughout the book several new ensemble classification methods based on fuzzy logic and the rough set theory [20] will be introduced.

Fuzzy logic, since its introduction in 1965 [32], has been used in various areas of science, economics, manufacturing, medicine etc. It constitutes, along with other methods like neural networks or evolutionary programming, the idea of soft computing. Neuro-fuzzy systems (NFS) [10, 15, 22, 23] are synergistic fusion of fuzzy logic and neural networks and they combine the best features of both approaches:

- the knowledge in the form of fuzzy rules reflects the natural language description of the problem,
- it is easy to achieve high accuracy thanks to learning inherited from neural networks.

R. Scherer: Multiple Fuzzy Classification Systems, STUDFUZZ 288, pp. 1–5.
springerlink.com

The intelligibility of the rules depends mainly on the conditional part of the fuzzy rules and the learning algorithm should be designed properly to retain it. Generally fuzzy systems can be divided, depending on the connective between the antecedent and the consequent in fuzzy rules, as follows:

i. Takagi-Sugeno method - consequents are functions of inputs,
ii. Mamdani-type reasoning method - consequents and antecedents are related by the min operator or generally by a t-norm,
iii. Logical-type reasoning method - consequents and antecedents are related by fuzzy implications, e.g. binary, Łukasiewicz, Zadeh etc.

Another approach is a fuzzy relational model [6, 21] where we define all possible connections between input and output linguistic terms. An advantage of this approach is great flexibility of the system. Input and output terms are fully interconnected. Moreover, the connections can be modeled by changing the elements of the relation matrix. The relation matrix can be regarded as a set of elements similar to rule weights in classical fuzzy systems [9, 16]. Relational fuzzy systems were used successfully to e.g. control [5] and classification tasks [1, 28, 30]. This book presents some applications of relational neuro-fuzzy systems [25, 27, 26]. Such neural network-like structures allow to use more scenarios than in the case of ordinary relational structures. For example, we can set fuzzy linguistic values in advance and then fine-tune the model mapping by changing relation elements using gradient learning. Gradient learning is an important advantage of relational neuro-fuzzy systems comparing to original fuzzy relational systems.

1.1 A Brief Summary of the Content

In the book we build ensembles of various types of neuro-fuzzy classifiers . A serious drawback of fuzzy system boosting ensembles is that such ensembles contain separate rule bases which cannot be directly merged. As systems are separate, we cannot treat fuzzy rules coming from different systems as rules from the same (single) system. In the book, this problem is addressed by a novel design of fuzzy systems constituting the ensemble, resulting in normalization of individual rule bases during learning. To our best knowledge such an approach has not yet been presented in the literature.

The next important problem raised in the book are ensembles that are able to work with missing data. They are based on rough-neuro-fuzzy systems [17, 18]. We derived several architectures of such ensembles.

The book is organized as follows. Chapter 2 presents some elements of fuzzy logic and fuzzy inference and Chapter 3 selected ensemble techniques used in the book. Chapter 4 concerns above–mentioned relational neuro-fuzzy systems and methods to join them into larger ensembles. Chapters 5–7 present ensembles consisted of, respectively, Mamdani, logical and Takagi-Sugeno neuro-fuzzy systems. Rough-neuro-fuzzy systems used to build boosting ensembles are presented in Chapter 8. To sum up, the book contributes with the following new methods:

- New method of backpropagation learning that takes into account boosting weights.
- Modification of clustering algorithm for ensembles.
- Novel design of the Mamdani fuzzy systems constituting the ensemble, resulting in normalization of individual rule bases during learning.
- New ensembles of logical type neuro-fuzzy systems.
- New flexible relational neuro-fuzzy systems.
- Novel ensemble methods for relational neuro-fuzzy systems.
- New types of classification ensembles of the Takagi-Sugeno fuzzy systems.
- A family of various rough-neuro-fuzzy ensembles.

Novel and original methods of learning suited for ensembles and new methods of creating ensembles are presented in Sections 3.4 and 3.5 and chapters 4–8.

1.2 Datasets Used in Experiments

Throughout this book several datasets have been used. They were obtained from the UC Irvine Machine Learning Repository [8].

1.2.1 Glass Identification

The Glass Identification problem [8] consists in an identification of glass type on the basis of 9 attributes for forensic purposes. The dataset has 214 instances and each instance is described by nine attributes (RI: refractive index, Na: sodium, Mg: magnesium, Al: aluminium, Si: silicon, K: potassium, Ca: calcium, Ba: barium, Fe: iron). All attributes are continuous. There are two classes: the window glass and the non-window glass. The goal is to classify 214 instances of glass into window and non-window glass basing on 9 numeric features.

1.2.2 Ionosphere

Ionosphere problem consists of radar data collected by a system in Goose Bay, Labrador [8]. System of 16 high-frequency antennas targeted on free electrons in the ionosphere returned either evidence or no evidence of some type of structure in the ionosphere. Each object is described by 34 continuous attributes and belongs to one of two classes. The data were divided randomly into 246 learning and 105 testing instances.

1.2.3 Pima Indians Diabetes Problem

The Pima Indians diabetes (PID) [8] data contains two classes, eight attributes (number of times pregnant, plasma glucose concentration in an oral glucose tolerance test, diastolic blood pressure (mm Hg), triceps skin fold thickness (mm), 2-hour serum insulin (mu U/ml), body mass index (weight in kg/(height in m)2), diabetes pedigree function, age (years)). We consider 768 instances, 500 (65.1%) healthy and 268 (34.9%) diabetes cases. All patients were females at least 21 years old of Pima Indian heritage, living near Phoenix, Arizona. It should be noted that about 33% of this population suffers from diabetes. In our experiments, all sets are divided into a learning sequence (576 sets) and a testing sequence (192 sets).

1.2.4 Wisconsin Breast Cancer Database

Wisconsin Breast Cancer Database consists of 699 instances of binary classes (benign or malignant type of cancer). Classification is based on 9 features (clump thickness, uniformity of cell size, uniformity of cell shape, marginal adhesion, single epithelial cell size, bare nuclei, bland chromatin, normal nucleoli, mitoses). From the data set 205 instances were taken into testing data and 16 instances with missing features were removed.

References

1. Babuska, R.: Fuzzy Modeling For Control. Kluwer Academic Press, Boston (1998)
2. Bezdek, J., Keller, J., Krisnapuram, R., Pal, N.: Fuzzy Models and Algorithms for Pattern Recognition and Image Processing. Kluwer Academic Press (1999)
3. Bezdek, J.C., Pal, S.: Fuzzy Models for Pattern Recognition. IEEE Press, New York (1992)
4. Bishop, C.M.: Neural Networks for Pattern recognition. Clarendon Press, Oxford (1995)
5. Branco, P., Dente, J.: A fuzzy relational identification algorithm and its application to predict the behaviour of a motor drive system. Fuzzy Sets and Systems 109, 343–354 (2000)
6. Ciaramella, A., Tagliaferri, R., Pedrycz, W., Nola, A.D.: Fuzzy Relational Neural Network for Data Analysis. In: Di Gesú, V., Masulli, F., Petrosino, A. (eds.) WILF 2003. LNCS (LNAI), vol. 2955, pp. 103–109. Springer, Heidelberg (2006)
7. Duda, R.O., Hart, P.E., Stork, D.G.: Pattern Classification. Wiley-Interscience Publication (2000)
8. Frank, A., Asuncion, A.: UCI machine learning repository (2010), `http://archive.ics.uci.edu/ml`
9. Ishibuchi, H., Nakashima, T.: Effect of rule weights in fuzzy rule-based classification systems. IEEE Trans. on Fuzzy Systems 9, 506–515 (2000)
10. Jang, R.J.S., Sun, C.T., Mizutani, E.: Neuro-Fuzzy and Soft Computing, A Computational Approach to Learning and Machine Intelligence. Prentice-Hall, Upper Saddle River (1997)
11. Kuncheva, L.: Fuzzy Classifier Design. Physica-Verlag, Heidelberg (2000)
12. Kuncheva, L.: Combining Pattern Classifiers. STUDFUZZ. John Wiley & Sons (2004)

13. Manning, C.D., Raghavan, P., Schütze, H.: Introduction to Information Retrieval. Cambridge University Press (2008)
14. Meir, R., Rätsch, G.: An Introduction to Boosting and Leveraging. In: Mendelson, S., Smola, A.J. (eds.) Advanced Lectures on Machine Learning. LNCS (LNAI), vol. 2600, pp. 118–183. Springer, Heidelberg (2003)
15. Nauck, D.: Foundations of Neuro-Fuzzy Systems. John Wiley, Chichester (1997)
16. Nauck, D., Kruse, R.: How the learning of rule weights affects the interpretability of fuzzy systems. In: Proceedings of 1998 IEEE World Congress on Computational Intelligence, FUZZ-IEEE, pp. 1235–1240 (1998)
17. Nowicki, R.: On combining neuro-fuzzy architectures with the rough set theory to solve classification problems with incomplete data. IEEE Trans. Knowl. Data Eng. 20(9), 1239–1253 (2008), doi:10.1109/TKDE.2008.64
18. Nowicki, R.: Rough-neuro-fuzzy structures for classification with missing data. IEEE Trans. Syst., Man, Cybern. B 39, 1334–1347 (2009)
19. Nowicki, R., Scherer, R.: A hierarchical fuzzy system with fuzzy intermediate variables. In: Proceedings of The 9th Zittau Fuzzy Colloquium 2001, Germany, pp. 88–93 (2001)
20. Pawlak, Z.: Rough Sets: Theoretical Aspects of Reasoning About Data. Kluwer, Dordrecht (1991)
21. Pedrycz, W.: Fuzzy Control and Fuzzy Systems. Research Studies Press, London (1989)
22. Pedrycz, W., Gomide, F.: An Introduction to Fuzzy Sets, Analysis and Design. The MIT Press, Cambridge (1998)
23. Rutkowski, L.: Flexible Neuro-Fuzzy Systems. Kluwer Academic Publishers (2004)
24. Schapire, R.E.: A brief introduction to boosting. In: Conference on Artificial Intelligence, pp. 1401–1406 (1999)
25. Scherer, R.: Neuro-fuzzy relational systems for nonlinear approximation and prediction. Nonlinear Analysis 71, e1420–e1425 (2009), `http://linkinghub.elsevier.com/retrieve/pii/S0362546X09001898`, doi:10.1016/j.na.2009.01.18
26. Scherer, R., Rutkowski, L.: Neuro-fuzzy relational systems. In: 2002 International Conference on Fuzzy Systems and Knowledge Discovery, Singapore, November 18-22, pp. 44–48 (2002)
27. Scherer, R., Rutkowski, L.: Relational equations initializing neuro-fuzzy system. In: 10th Zittau Fuzzy Colloquium, 2002, Zittau, Germany, pp. 212–217 (2002)
28. Scherer, R., Rutkowski, L.: Neuro-Fuzzy Relational Classifiers. In: Rutkowski, L., Siekmann, J.H., Tadeusiewicz, R., Zadeh, L.A. (eds.) ICAISC 2004. LNCS (LNAI), vol. 3070, pp. 376–380. Springer, Heidelberg (2004)
29. Scherer, R., Starczewski, J., Gaweda, A.: New Methods for Uncertainty Representations in Neuro-Fuzzy Systems. In: Wyrzykowski, R., Dongarra, J., Paprzycki, M., Waśniewski, J. (eds.) PPAM 2004. LNCS, vol. 3019, pp. 659–667. Springer, Heidelberg (2004)
30. Setness, M., Babuska, R.: Fuzzy relational classifier trained by fuzzy clustering. IEEE Transactions on Systems, Man and Cybernetics - Part B: Cybernetics 29(5), 619–625 (1999)
31. Starczewski, J., Scherer, R., Korytkowski, M., Nowicki, R.: Modular Type-2 Neuro-fuzzy Systems. In: Wyrzykowski, R., Dongarra, J., Karczewski, K., Wasniewski, J. (eds.) PPAM 2007. LNCS, vol. 4967, pp. 570–578. Springer, Heidelberg (2008), `http://www.springerlink.com/content/c412v2479u38033h/`
32. Zadeh, L.A.: Fuzzy sets. Information Control 8, 338–353 (1965)

Chapter 2
Introduction to Fuzzy Systems

Fuzzy logic since its conception in 1965 [11] has been used in various areas of science, economics, manufacturing, medicine etc [1, 2, 3, 4, 7, 8, 9]. It constitutes, along with other methods such as neural networks or evolutionary algorithms, the idea of soft computing [5]. Fuzzy sets used in fuzzy rules can be a tool to model linguistic values like "small" or "high" [12]. This chapter presents basic definitions of fuzzy logic and fuzzy systems based on [10].

2.1 Elements of the Theory of Fuzzy Sets

Fuzzy sets can be used to define imprecise and ambiguous expressions such as "high tree" or "fast car". Fuzzy sets are defined in the so-called universe of discourse, denoted by $\mathbf{X}$, which is the range of possible elements of the fuzzy set.

Definition 2.1. The fuzzy set A in a given (non-empty) space $\mathbf{X}$, which is denoted as $A \subseteq \mathbf{X}$, is the set of pairs

$$A = \{(x, \mu_A(x)); x \in \mathbf{X}\}, \tag{2.1}$$

in which

$$\mu_A : \mathbf{X} \rightarrow [0, 1] \tag{2.2}$$

is the membership function of a fuzzy set A. This function assigns to each element $x \in \mathbf{X}$ its membership degree to the fuzzy set A, and we can distinguish 3 cases:

1) $\mu_A(x) = 1$ means the full membership of element x to the fuzzy set A, i.e. $x \in A$,
2) $\mu_A(x) = 0$ means the lack of membership of element x to the fuzzy set A, i.e. $x \notin A$,
3) $0 < \mu_A(x) < 1$ means a partial membership of element x to the fuzzy set A.

In the literature some authors use symbolic notations of fuzzy sets. If $\mathbf{X}$ is a space with a finite number of elements, $\mathbf{X} = \{x_1, ..., x_n\}$, then the fuzzy set $A \subseteq \mathbf{X}$ is denoted as

R. Scherer: Multiple Fuzzy Classification Systems, STUDFUZZ 288, pp. 7–28.
springerlink.com

$$A = \frac{\mu_A(x_1)}{x_1} + \frac{\mu_A(x_2)}{x_2} + \ldots + \frac{\mu_A(x_n)}{x_n} = \sum_{i=1}^{n} \frac{\mu_A(x_i)}{x_i}. \tag{2.3}$$

It should be reminded that the elements $x_i \in \mathbf{X}$ may be not only numbers, but also persons, objects or other notions. Notation (2.3) has a symbolic character. The fraction bar does not symbolize the division, but means the assigning membership degrees $\mu_A(x_1), \ldots, \mu_A(x_n)$ to particular elements $x_1, \ldots, x_n$. In other words, the notation

$$\frac{\mu_A(x_i)}{x_i} \quad i = 1, \ldots, n \tag{2.4}$$

denotes the pair

$$(x_i, \mu_A(x_i)) \quad i = 1, \ldots, n. \tag{2.5}$$

Similarly, the "+" sign does not mean the addition, but the union of sets (2.5). It is worth noting that the non-fuzzy sets may be notated symbolically in a similar convention. For example, the set of school grades shall be symbolically noted as

$$D = 1 + 2 + 3 + 4 + 5 + 6, \tag{2.6}$$

which is equal to the notation

$$D = \{1, 2, 3, 4, 5, 6\}. \tag{2.7}$$

If $\mathbf{X}$ is a space with an infinite number of elements, then the fuzzy set $A \subseteq \mathbf{X}$ is notated symbolically as

$$A = \int_{\mathbf{X}} \frac{\mu_A(x)}{x}. \tag{2.8}$$

In some applications, the standard forms of membership function are used. Below we shall specify these functions and will present their graphic representations.

1. The *singleton* function shall be defined as follows:

$$\mu_A(x) = \begin{cases} 1, \text{ if } x = \overline{x}, \\ 0, \text{ if } x \neq \overline{x}. \end{cases} \tag{2.9}$$

The singleton is a specific membership function, as it takes the value 1 only in a single point of the universe of discussion, belonging fully to the fuzzy set. In other points, it takes the value of 0. This membership function characterizes a single-element fuzzy set. The only element having the full membership to the fuzzy set A is the point $\overline{x}$. The singleton membership function is mainly used to perform fuzzification operation applied in fuzzy inference systems.

2. *Gaussian* membership function (Fig. 2.1 b) is described by the formula

$$\mu_A(x) = \exp\left(-\left(\frac{x-\overline{x}}{\sigma}\right)^2\right), \tag{2.10}$$

where $\bar{x}$ is the middle and σ defines the width of the Gaussian curve. It is the most common membership function.
3. *Bell* membership function (Fig. 2.1 c) takes the form of

$$\mu(x;a,b,c) = \frac{1}{1+\left|\frac{x-c}{a}\right|^{2b}}, \tag{2.11}$$

where the parameter a defines its width, the parameter b its slopes, and the parameter c its center.

4. Membership function of *class s* (Fig. 2.1 d) is defined as

$$s(x;a,b,c) = \begin{cases} 0 & \text{for } x \le a, \\ 2\left(\frac{x-a}{c-a}\right)^2 & \text{for } a < x \le b, \\ 1-2\left(\frac{x-c}{c-a}\right)^2 & \text{for } b < x \le c, \\ 1 & \text{for } x > c. \end{cases} \tag{2.12}$$

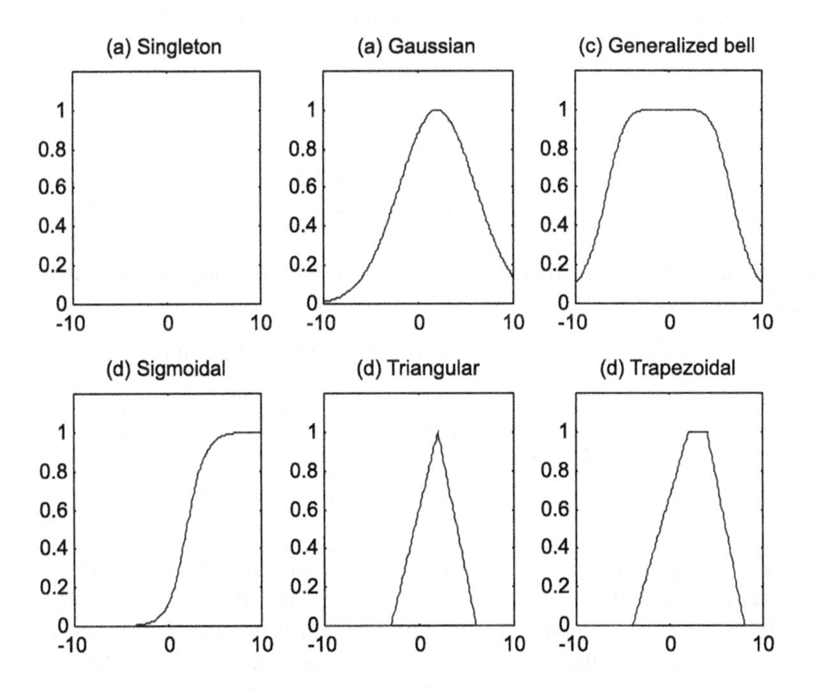

Fig. 2.1 Examples of membership functions of one dimensional fuzzy sets [10]

where $b = (a+c)/2$. The membership function belonging to this class takes a graphic form resembling the letter "s", and its shape depends on the selection $x = b = (a+c)/2$ the membership function of class s takes the value of 0.5.

5. The membership function of *class* π is defined by the membership function of class s

$$\pi(x;b,c) = \begin{cases} s(x;c-b,\ c-b/2,c) & \text{for } x \leq c, \\ 1-s(x;c,c+b/2,c+b) & \text{for } x > c. \end{cases} \tag{2.13}$$

The membership function of class π takes the zero values for $x \geq c+b$ and $x \leq c-b$. In points $x = c \pm b/2$ its value is 0.5. The membership function of class π takes the zero values for $x \geq c+b$ and $x \leq c-b$. In points $x = c \pm b/2$ its value is 0.5.

6. The membership function of *class* γ is given by the formula

$$\gamma(x;a,b) = \begin{cases} 0 & \text{for } x \leq a, \\ \dfrac{x-a}{b-a} & \text{for } a < x \leq b, \\ 1 & \text{for } a > b. \end{cases} \tag{2.14}$$

The Reader will easily notice the analogies between the shapes of the membership function of class s and γ.

7. The membership function of *class* t is defined as follows:

$$t(x;a,b,c) = \begin{cases} 0 & \text{for } x \leq a, \\ \dfrac{x-a}{b-a} & \text{for } a < x \leq b, \\ \dfrac{c-x}{c-b} & \text{for } b < x \leq c, \\ 0 & \text{for } x > c. \end{cases} \tag{2.15}$$

In some applications, the membership function of class t may be alternative to the function of class π.

8. The membership function of *class* L is defined by the formula

$$L(x;a,b) = \begin{cases} 1 & \text{for } x \leq a, \\ \dfrac{b-x}{b-a} & \text{for } a < x \leq b, \\ 0 & \text{for } a > b. \end{cases} \tag{2.16}$$

Above, we provided some examples of standard membership function for fuzzy sets defined in the space of real numbers i.e. $\mathbf{X} \subset \mathbf{R}$. When $\mathbf{X} \subset \mathbf{R}^n$, $\mathbf{x} = [x_1, ..., x_n]^T$, $n > 1$, we may distinguish two cases. The first case occurs when we assume the independence of particular variables x_i, $i = 1, ..., n$. Then the multidimensional membership functions are created by applying the definition of Cartesian product of fuzzy sets (Definition 2.11) and using standard membership functions of one variable. In case the variables x_i are dependent, we apply the multidimensional membership function. Below, three examples of such functions are specified:

1. The membership function of *class Π* (Fig. 2.2 a) is defined as follows:

$$\mu_A(\mathbf{x}) = \begin{cases} 1-2\cdot\left(\dfrac{\|\mathbf{x}-\bar{\mathbf{x}}\|}{\alpha}\right)^2 & \text{for } \|\mathbf{x}-\bar{\mathbf{x}}\| \le \dfrac{1}{2}\alpha, \\ 2\cdot\left(1-\dfrac{\|\mathbf{x}-\bar{x}\|}{\alpha}\right)^2 & \text{for } \dfrac{1}{2}\alpha < \|\mathbf{x}-\bar{\mathbf{x}}\| \le \alpha, \\ 0 & \text{for } \|\mathbf{x}-\bar{\mathbf{x}}\| > \alpha, \end{cases} \tag{2.17}$$

when $\bar{\mathbf{x}}$ is the center of the membership function, $\alpha > 0$ is the parameter defining its spread.

2. The radial membership function (Fig. 2.2 b) takes the form

$$\mu_A(\mathbf{x}) = \mathrm{e}^{\frac{\|\mathbf{x}-\bar{\mathbf{x}}\|^2}{2\cdot\sigma^2}}, \tag{2.18}$$

where $\bar{x}$ is the center, and the value of the parameter σ influences the shape of this function.

3. The ellipsoidal membership function (Fig. 2.2 c) is defined as follows:

$$\mu_A(\mathbf{x}) = \exp\left(-\frac{(\mathbf{x}-\bar{\mathbf{x}})^T \mathbf{Q}^{-1}(\mathbf{x}-\bar{\mathbf{x}})}{\alpha}\right), \tag{2.19}$$

where $\bar{\mathbf{x}}$ is the center, $\alpha > 0$ is the parameter defining the spread of this function, and $\mathbf{Q}$ is the so-called covariance matrix. By modifying this matrix, we may model the shape of this function.

Definition 2.2. The set of elements of the universe $\mathbf{X}$, for which $\mu_A(x) > 0$, is called *the support of a fuzzy set* A and is denoted as supp A (*support*), i.e.

$$\operatorname{supp} A = \{x \in \mathbf{X};\ \mu_A(x) > 0\}. \tag{2.20}$$

Definition 2.3. The height of a fuzzy set A $h(A)$ is defined as

$$h(A) = \sup_{x\in\mathbf{X}} \mu_A(x). \tag{2.21}$$

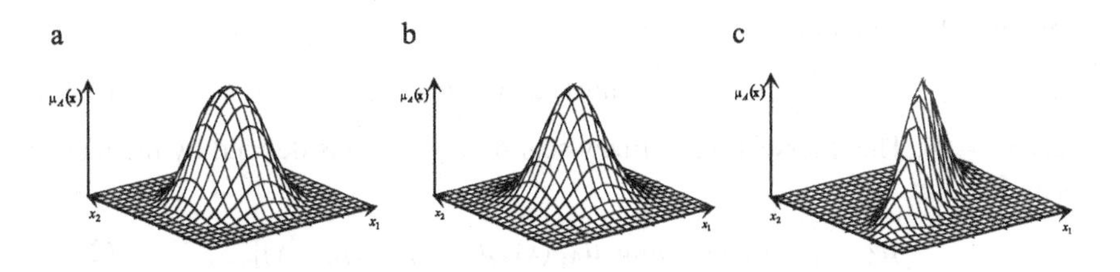

Fig. 2.2 Examples of membership functions of multidimensional fuzzy sets [10]

Definition 2.4. The fuzzy set A is called empty which shall be notated $A = \varnothing$ if and only if $\mu_A(x) = 0$ for each $x \in \mathbf{X}$.

Definition 2.5. The fuzzy set A is *included* in the fuzzy set B, which shall be notated $A \subset B$, if and only if

$$\mu_A(x) \leq \mu_B(x) \tag{2.22}$$

for each $x \in \mathbf{X}$.

Definition 2.6. The fuzzy set A is *equal* to the fuzzy set B, which shall be notated $A = B$, if and only if

$$\mu_A(x) = \mu_B(x) \tag{2.23}$$

for each $x \in \mathbf{X}$.

The above definition, similarly to Definition 2.5, is not "flexible", as it does not consider the case when the values of the membership functions $\mu_A(x)$ and $\mu_B(x)$ are almost equal. Then we can introduce the term of equality degree of fuzzy sets A and B as for example

$$E(A = B) = 1 - \max_{x \in T} |\mu_A(x) - \mu_B(x)|, \tag{2.24}$$

where $T = \{x \in X : \mu_A(x) \neq \mu_B(x)\}$. There exist many definitions of the inclusion degree and the equality degree.

Definition 2.7. *α-cut* of the fuzzy set $A \subseteq \mathbf{X}$, notated as A_α is called the following non-fuzzy set:

$$A_\alpha = \{x \in \mathbf{X} : \mu_A(x) \geq \alpha\}, \quad \forall_{\alpha \in [0,1]}, \tag{2.25}$$

or the set defined by the characteristic function

$$\chi_{A_\alpha}(x) = \begin{cases} 1 \text{ for } \mu_A(x) \geq \alpha, \\ 0 \text{ for } \mu_A(x) < \alpha. \end{cases} \tag{2.26}$$

$$\alpha_2 < \alpha_1 \implies A_{\alpha_1} \subset A_{\alpha_2}. \tag{2.27}$$

2.2 Operations on Fuzzy Sets

This section presents some operations on fuzzy sets.

Definition 2.8. The intersection of fuzzy sets A, $B \subseteq \mathbf{X}$ is the fuzzy set $A \cap B$ with the membership function

$$\mu_{A \cap B}(x) = \min(\mu_A(x), \mu_B(x)) \tag{2.28}$$

for each $x \in \mathbf{X}$. The intersection of fuzzy sets $A_1, A_2, ..., A_n$ is defined by the membership function

$$\mu_{A_1 \cap A_2 \cdots \cap A_n}(x) = \min[\mu_{A_1}(x), \mu_{A_2}(x), ..., \mu_{A_n}(x)] \tag{2.29}$$

for each $x \in \mathbf{X}$.

Definition 2.9. The union of fuzzy sets A and B is the fuzzy set $A \cup B$ defined by the membership function

$$\mu_{A \cup B}(x) = \max\left(\mu_A(x), \mu_B(x)\right) \tag{2.30}$$

for each $x \in \mathbf{X}$. The membership function of the union of fuzzy sets $A_1, A_2, ..., A_n$ is expressed by the formula

$$\mu_{A_1 \cup A_2 \cup \cdots \cup A_n}(x) = \max\left[\mu_{A_1}(x), \mu_{A_2}(x), ..., \mu_{A_n}(x)\right] \tag{2.31}$$

for each $x \in \mathbf{X}$.

Definition 2.10. The complement of a fuzzy set $A \subseteq \mathbf{X}$ is the fuzzy set $\widehat{A}$ with the membership function

$$\mu_{\widehat{A}}(x) = 1 - \mu_A(x) \tag{2.32}$$

for each $x \in \mathbf{X}$.

Definition 2.11. The Cartesian product of fuzzy sets $A \subseteq \mathbf{X}$ and $B \subseteq \mathbf{Y}$ is notated as $A \times B$ and defined as

$$\mu_{A \times B}(x, y) = \min\left(\mu_A(x), \mu_B(y)\right) \tag{2.33}$$

or

$$\mu_{A \times B}(x, y) = \mu_A(x)\, \mu_B(y) \tag{2.34}$$

for each $x \in \mathbf{X}$ and $y \in \mathbf{Y}$. The Cartesian product of fuzzy sets $A_1 \subseteq \mathbf{X}_1, A_2 \subseteq \mathbf{X}_2, ..., A_n \subseteq \mathbf{X}_n$ is notated as $A_1 \times A_2 \times ... \times A_n$ and defined as

$$\mu_{A_1 \times A_2 \times ... \times A_n}(x_1, x_2, ..., x_n) = \min(\mu_{A_1}(x_1), \mu_{A_2}(x_2), ..., \mu_{A_n}(x_n)) \tag{2.35}$$

or

$$\mu_{A_1 \times A_2 \times ... \times A_n}(x_1, x_2, ..., x_n) = \mu_{A_1}(x_1)\, \mu_{A_2}(x_2), ..., \mu_{A_n}(x_n) \tag{2.36}$$

for each $x_1 \in \mathbf{X}_1$, $x_2 \in \mathbf{X}_2$, ..., $x_n \in \mathbf{X}_n$.

Definition 2.12. The concentration of a fuzzy set $A \subseteq \mathbf{X}$ shall be notated as $CON(A)$ and defined as

$$\mu_{CON(A)}(x) = \left(\mu_A(x)\right)^2 \tag{2.37}$$

for each $x \in \mathbf{X}$.

Definition 2.13. The dilation of a fuzzy set $A \subseteq \mathbf{X}$ shall be notated as $DIL(A)$ and defined as

$$\mu_{DIL(A)}(x) = \left(\mu_A(x)\right)^{0.5} \tag{2.38}$$

for each $x \in \mathbf{X}$.

2.3 Triangular Norms and Negations

Earlier in this chapter, we have defined the operations of intersection and union of fuzzy sets as

$$\mu_{A\cap B}(x) = \min\left(\mu_A(x), \mu_B(x)\right),$$
$$\mu_{A\cup B}(x) = \max\left(\mu_A(x), \mu_B(x)\right).$$

The intersection of fuzzy sets can be defined more generally as

$$\mu_{A\cap B}(x) = T\left(\mu_A(x), \mu_B(x)\right), \tag{2.39}$$

where the function T is the so-called t-norm. Therefore, $\min(\mu_A(x), \mu_B(x)) = T(\mu_A(x), \mu_B(x))$ is an example of operation of the t-norm. Similarly, the union of fuzzy sets is defined as follows:

$$\mu_{A\cup B}(x) = S\left(\mu_A(x), \mu_B(x)\right), \tag{2.40}$$

where the function S is the so-called t-conorm. In this case, $\max(\mu_A(x), \mu_B(x)) = S(\mu_A(x), \mu_B(x))$ is an example of the t-conorm. It is worth noting that the t-norms and the t-conorms belong to the so-called triangular norms. Below formal definitions will be presented.

Definition 2.14. The function of two variables T

$$T : [0,1] \times [0,1] \rightarrow [0,1] \tag{2.41}$$

is called a t-norm, if
(i) function T is nondecreasing with relation to both arguments

$$T(a,c) \leq T(b,d) \quad \text{for} \quad a \leq b,\ c \leq d \tag{2.42}$$

(ii) function T satisfies the condition of commutativity

$$T(a,b) = T(b,a) \tag{2.43}$$

(iii) function T satisfies the condition of associativity

$$T(T(a,b),c) = T(a,T(b,c)) \tag{2.44}$$

(iv) function T satisfies the boundary condition

$$T(a,1) = a, \tag{2.45}$$

where $a,b,c,d \in [0,1]$.

From the assumptions it follows that

$$T(a,0) = T(0,a) \leq T(0,1) = 0. \tag{2.46}$$

Therefore, the second boundary condition takes the form

$$T(a,0) = 0. \tag{2.47}$$

Further in this chapter, we shall notate the operation of t-norm on arguments a and b in the following way

$$T(a,b) = a \overset{T}{*} b. \tag{2.48}$$

If, for instance, a and b are identified with the membership functions of fuzzy sets A and B, then we write equality (2.39) as

$$\mu_{A \cap B}(x) = T(\mu_A(x), \mu_B(x)) = \mu_A(x) \overset{T}{*} \mu_A(x). \tag{2.49}$$

Using property (2.44), the definition of t-norm may be generalized for the case of a t-norm of multiple variables

$$\begin{aligned} \overset{n}{\underset{i=1}{T}} \{a_i\} &= T\left\{ \overset{n-1}{\underset{i=1}{T}} \{a_i\}, a_n \right\} = T\{a_1, a_2, ..., a_n\} = T\{\mathbf{a}\} \\ &= a_1 \overset{T}{*} a_2 \overset{T}{*} \ldots \overset{T}{*} a_n. \end{aligned} \tag{2.50}$$

Definition 2.15. The function of two variables S

$$S : [0,1] \times [0,1] \to [0,1] \tag{2.51}$$

is called a t-conorm, if it is nondecreasing with relation to both arguments, meets the condition of commutativity and associativity, and the following boundary condition is met:

$$S(a,0) = a. \tag{2.52}$$

From the assumptions and condition (2.52) we get:

$$S(a,1) = S(1,a) \geq S(1,0) = 1. \tag{2.53}$$

Therefore, the second boundary condition takes the form

$$S(a,1) = 1. \tag{2.54}$$

The operation of t-conorm on arguments a and b will be notated in the following way

$$S(a,b) = a \overset{S}{*} b. \tag{2.55}$$

Using the property of associativity, the above definition may be generalized for the case of a t-conorm of multiple variables

$$\begin{aligned} \overset{n}{\underset{i=1}{S}} \{a_i\} &= S\left\{ \overset{n-1}{\underset{i=1}{S}} \{a_i\}, a_n \right\} = S\{a_1, a_2, ..., a_n\} = S\{\mathbf{a}\} \\ &= a_1 \overset{S}{*} a_2 \overset{S}{*} \ldots \overset{S}{*} a_n. \end{aligned} \tag{2.56}$$

The most popular triangular norms are of the min/max type, described by the following formulas

$$T_M\{a_1,a_2\} = \min\{a_1,a_2\}, \tag{2.57}$$

$$S_M\{a_1,a_2\} = \max\{a_1,a_2\}, \tag{2.58}$$

$$T_M\{a_1,a_2,...,a_n\} = \min_{i=1,...,n}\{a_i\}, \tag{2.59}$$

$$S_M\{a_1,a_2,...,a_n\} = \max_{i=1,...,n}\{a_i\}. \tag{2.60}$$

2.4 Fuzzy Relations and Their Properties

Fuzzy relations allow to define relationships between fuzzy sets. Below we present the definitions of fuzzy relation and composition of fuzzy relations [10].

Definition 2.16. The fuzzy relation R between two non-empty (non-fuzzy) sets $\mathbf{X}$ and $\mathbf{Y}$ is called the fuzzy set determined on the Cartesian product $\mathbf{X}\times\mathbf{Y}$, i.e.

$$R \subseteq \mathbf{X}\times\mathbf{Y} = \{(x,y) : x\in\mathbf{X},\ y\in\mathbf{Y}\}. \tag{2.61}$$

In other words, the fuzzy relation is a set of pairs

$$R = \{((x,y),\mu_R(x,y))\}, \quad \forall_{x\in\mathbf{X}}\forall_{y\in\mathbf{Y}}, \tag{2.62}$$

where

$$\mu_R : \mathbf{X}\times\mathbf{Y}\rightarrow[0,1] \tag{2.63}$$

is the membership function. To each pair (x,y), $x\in\mathbf{X}$, $y\in\mathbf{Y}$ this function assigns membership degree $\mu_R(x,y)$, which is interpreted as strength of relation between elements $x\in\mathbf{X}$ and $y\in\mathbf{Y}$.

In the fuzzy sets theory, an important role is given to the notion of composition of two fuzzy relations. Let us consider three non-fuzzy sets $\mathbf{X}$, $\mathbf{Y}$, $\mathbf{Z}$ and two fuzzy relations $R\subseteq\mathbf{X}\times\mathbf{Y}$ and $S\subseteq\mathbf{Y}\times\mathbf{Z}$ with membership functions $\mu_R(x,y)$ and $\mu_S(y,z)$, respectively.

Definition 2.17. Composition of sup-T type of fuzzy relations $R\subseteq\mathbf{X}\times\mathbf{Y}$ and $S\subseteq\mathbf{Y}\times\mathbf{Z}$ is called a fuzzy relation $R\circ S\subseteq\mathbf{X}\times\mathbf{Z}$ with the membership function

$$\mu_{R\circ S}(x,z) = \sup_{y\in\mathbf{Y}}\left\{\mu_R(x,y)\overset{T}{*}\mu_S(y,z)\right\}. \tag{2.64}$$

A specific form of the membership function $\mu_{R\circ S}(x,z)$ of the composition $R\circ S$ depends on the adopted t-norm in formula (2.64). If we take min as a t-norm, i.e. $T(a,b)=\min(a,b)$, then equation (2.64) may be notated as follows:

$$\mu_{R\circ S}(x,z) = \sup_{y\in\mathbf{Y}} \{\min[\mu_R(x,y), \mu_S(y,z)]\}. \tag{2.65}$$

Formula (2.65) is known in the literature as the sup-min type composition. If the set **Y** has an infinite number of elements, then the composition sup-min comes down to the max-min type composition of the form

$$\mu_{R\circ S}(x,z) = \max_{y\in\mathbf{Y}} \{\min[\mu_R(x,y), \mu_S(y,z)]\}. \tag{2.66}$$

As we have already mentioned, the composition of a fuzzy set with a fuzzy relation is particularly important. The composition of this type will be used repeatedly further in this chapter. Let us consider a fuzzy set $A \subseteq \mathbf{X}$ and a fuzzy relation $R \subseteq \mathbf{X} \times \mathbf{Y}$ with membership functions $\mu_A(x)$ and $\mu_R(x,y)$, respectively.

Definition 2.18. Composition of a fuzzy set $A \subseteq \mathbf{X}$ and a fuzzy relation $R \subseteq \mathbf{X} \times \mathbf{Y}$ shall be notated $A \circ R$ and defined as a fuzzy set $B \subseteq \mathbf{Y}$

$$B = A \circ R \tag{2.67}$$

with the membership function

$$\mu_B(y) = \sup_{x\in\mathbf{X}} \left\{ \mu_A(x) \overset{T}{*} \mu_R(x,y) \right\}. \tag{2.68}$$

A specific form of formula (2.68) depends on the chosen t-norm and on properties of set **X**. Below, we shall present 4 cases:

1) If $T(a,b) = \min(a,b)$, then we obtain a composition of the sup-min type

$$\mu_B(y) = \sup_{x\in\mathbf{X}} \{\min[\mu_A(x), \mu_R(x,y)]\}. \tag{2.69}$$

2) If $T(a,b) = \min(a,b)$, and **X** is a set with a finite number of elements, then we obtain the composition of the max-min type

$$\mu_B(y) = \max_{x\in\mathbf{X}} \{\min[\mu_A(x), \mu_R(x,y)]\}. \tag{2.70}$$

3) If $T(a,b) = a \cdot b$, then we obtain a composition of the sup-product type

$$\mu_B(y) = \sup_{x\in\mathbf{X}} \{\mu_A(x) \cdot \mu_R(x,y)\}. \tag{2.71}$$

4) If $T(a,b) = a \cdot b$, and **X** is a set with a finite number of elements, then we obtain the composition of the max-product type

$$\mu_B(y) = \max_{x\in\mathbf{X}} \{\mu_A(x) \cdot \mu_R(x,y)\}. \tag{2.72}$$

2.5 Approximate Reasoning

2.5.1 *Basic Rules of Inference in Binary Logic*

In traditional (binary) logic, we infer about the truth of some sentences based on the truth of some other sentences. This reasoning shall be notated in the form of a schema: above the horizontal line, we shall put all the sentences, on the basis of which we make the reasoning, below the line, we shall put the inference. The schema of correct inference has such property that if all sentences above the horizontal line are true, then also the sentence below the line is true, as true sentences may lead only to a true inference. In this point capital letters A and B symbolize the sentences rather than fuzzy sets. Let A and B be the sentences, while the notation $A = 1(B = 1)$ means that the logical value A (B) is true and the notation $A = 0(B = 0)$ means, that the logical value of the sentence $A(B)$ is falsehood. We shall present below two rules of inference used in binary logic.

Definition 2.19. The reasoning rule modus ponens is defined by the following reasoning schema:

Premise	A
Implication	$A \to B$
Inference	B

(2.73)

Definition 2.20. The reasoning rule modus tollens is defined by the following reasoning schema:

Premise	$\overline{B}$
Implication	$A \to B$
Inference	$\overline{A}$

(2.74)

2.5.2 *Basic Rules of Inference in Fuzzy Logic*

If the statements in the modus ponens (2.73) and modus tollens (2.74) rules are characterized by fuzzy sets, binary inference rules have to be redefined. In this way, we obtain a generalized modus ponens inference rule and a generalized modus tollens inference rule [10].

Definition 2.21. A generalized (fuzzy) modus ponens inference rule is defined by the following reasoning schema:

Premise	x is A'
Implication	**IF** x is A **THEN** y is B
Inference	y is B'

(2.75)

where $A, A' \subseteq \mathbf{X}$ and $B, B' \subseteq \mathbf{Y}$ are fuzzy sets, and x and y are so-called linguistic variables.

According to the above definition, linguistic variables are variables which take as values words or sentences uttered in the natural language. Examples may be provided by such statements as "low speed", "temperate temperature" or "young person". These statements may be formalized by assigning some fuzzy sets to them. It should be stressed that linguistic variables may, apart from word values, take numerical values just like ordinary mathematical variables.

Definition 2.22. A generalized (fuzzy) modus tollens inference rule is defined by the following reasoning schema:

Premise	y is B'
Implication	**IF** x is A **THEN** y is B
Inference	x is A'

(2.76)

where $A, A' \subseteq \mathbf{X}$ and $B, B' \subseteq \mathbf{Y}$ are fuzzy sets, and x and y are the linguistic variables.

2.5.3 *Inference Rules for the Mamdani Model*

In the case of the Mamdani model, the membership functions $\mu_{A \to B}(x, y)$ are determined as follows:

$$\mu_{A \to B}(x, y) = T\ (\mu_A(x), \mu_B(y)), \tag{2.77}$$

where T is any t-norm. We may interpret function T in formula (2.77) as the correlation function between the antecedens and consequences in fuzzy rules. Most often, the minimum type rule defined below is applied:

- **Minimum type rule**

$$\begin{aligned}\mu_{A \to B}(x, y) = \mu_R(x, y) &= \mu_A(x) \wedge \mu_B(y) \\ &= \min\left[\mu_A(x), \mu_B(y)\right].\end{aligned} \tag{2.78}$$

Another known rule is the product type rule (also referred to as the Larsen rule):

- **Product type rule (Larsen)**

$$\mu_{A\to B}(x,y) = \mu_R(x,y) = \mu_A(x)\cdot\mu_B(y). \tag{2.79}$$

Mamdani type rules are not implications in the logical meaning.

2.5.4 *Inference Rules for the Logical Model*

Inference in the logical fuzzy model is based on a function called fuzzy implication.

Definition 2.23. *A fuzzy implication* is the function $I : [0,1]^2 \to [0,1]$ meeting the following conditions:
a) if $a_1 \le a_3$, then $I(a_1,a_2) \ge I(a_3,a_2)$ for all $a_1,a_2,a_3 \in [0,1]$,
b) if $a_2 \le a_3$, then $I(a_1,a_2) \le I(a_1,a_3)$ for all $a_1,a_2,a_3 \in [0,1]$,
c) $I(0,a_2) = 1$ for all $a_2 \in [0,1]$,
d) $I(a_1,1) = 1$ for all $a_1 \in [0,1]$,
e) $I(1,0) = 0$.

Below membership functions $\mu_{A\to B}(x,y)$ for the logical inference are listed:

1. **Binary implication (Kleene-Dienes)**

$$\mu_{A\to B}(x,y) = \max\left[1-\mu_A(x),\mu_B(y)\right]. \tag{2.80}$$

2. **Łukaszewicz implication**

$$\mu_{A\to B}(x,y) = \min\left[1.1-\mu_A(x)+\mu_B(y)\right]. \tag{2.81}$$

3. **Reichenbach implication**

$$\mu_{A\to B}(x,y) = 1.1-\mu_A(x)+\mu_A(x)\cdot\mu_B(y). \tag{2.82}$$

4. **Fodor implication**

$$\mu_{A\to B}(x,y) = \begin{cases} 1, & \text{if}\mu_A(x) \le \mu_B(y), \\ \max\{1-\mu_A(x),\mu_B(y)\}, & \text{if}\mu_A(x) > \mu_B(y). \end{cases} \tag{2.83}$$

5. **Rescher implication**

$$\mu_{A\to B}(x,y) = \begin{cases} 1, \text{ if } \mu_A(x) \le \mu_B(y), \\ 0, \text{ if } \mu_A(x) > \mu_B(y). \end{cases} \tag{2.84}$$

6. **Goguen implication**

$$\mu_{A\to B}(x,y) = \begin{cases} \min\left[1,\dfrac{\mu_B(y)}{\mu_A(x)}\right], & \text{if } \mu_A(x) > 0, \\ 1, & \text{if } \mu_A(x) = 0. \end{cases} \tag{2.85}$$

7. **Gödel implication**

$$\mu_{A\to B}(x,y) = \begin{cases} 1, & \text{if } \mu_A(x) \le \mu_B(y), \\ \mu_B(y), & \text{if } \mu_A(x) > \mu_B(y). \end{cases} \tag{2.86}$$

8. **Yager implication**

$$\mu_{A\to B}(x,y) = \begin{cases} 1, & \text{if } \mu_A(x) = 0, \\ \mu_B(y)^{\mu_A(x)}, & \text{if } \mu_A(x) > 0. \end{cases} \tag{2.87}$$

9. **Zadeh implication**

$$\mu_{A\to B}(x,y) = \max\{\min[\mu_A(x), \mu_B(y)], 1-\mu_A(x)\}. \tag{2.88}$$

10. **Willmott implication**

$$\mu_{A\to B}(x,y) = \min\left\{\begin{array}{c} \max\{1-\mu_A(x),\ \mu_B(y)\}, \\ \max\{\mu_A(x), 1-\mu_B(y), \\ \min\{1-\mu_A(x), \mu_B(y)\}\}. \end{array}\right\} \tag{2.89}$$

11. **Dubois-Prade implication**

$$\mu_{A\to B}(x,y) = \begin{cases} 1-\mu_A(x), & \text{if } \mu_B(y) = 0 \\ \mu_B(y), & \text{if } \mu_A(x) = 1 \\ 1, & \text{otherwise} \end{cases} \tag{2.90}$$

Some of the above implications can be divided into the following groups [10]:
a) *S-implications* defined as follows:

$$I(a,b) = S\{1-a, b\}.$$

Examples of S-implication are implications 1, 2, 3 and 4. They meet all the conditions of Definition 2.23.
b) *R-implications* defined as follows:

$$I(a,b) = \sup_z \{z \mid T\{a,z\} \le b\}, \quad a,b \in [0,1].$$

Examples of R-implication are implications 6 and 7. They meet all the conditions of Definition 2.23.

c) *Q-implications* defined as follows:

$$I(a,b) = S\{N(a), T\{a,b\}\}, \quad a,b \in [0,1],$$

where $N(a)$ is a negation operator. An example of Q-implication is the Zadeh implication, which does not meet the condition a) and d) of Definition 2.23.

2.6 Fuzzy Inference Systems

In many issues concerning the technological processes control, it will be necessary to determine a model of the considered process. The knowledge of the model allows to select the appropriate controller. However, often it is very difficult to find an appropriate model, it is a problem which sometimes requires different simplifying assumptions. The application of the fuzzy sets theory to control technological processes does not require any knowledge of models of these processes. It is enough to formulate rules of procedure in the form of sentences like: **IF** ... **THEN**. Similarly, classification tasks may be solved. The approach using rules of the **IF** ... **THEN** type allows to solve a classification problem without the knowledge of probability densities of particular classes. Fuzzy control systems and classifiers are particular cases of fuzzy inference systems. Figure 2.3 illustrates a typical schema of such a system. It consists of the following elements:

1. rule base,
2. fuzzification block,
3. inference block,
4. defuzzification block.

Further in this chapter, we will assume that the inference block uses the Mamdani type model, in which the antecedents and consequents of rules are combined using a t-norm operation. Later in this book, we will present in detail the neuro-fuzzy structures built using both the Mamdani and the logical model. Now, we will discuss particular elements of the fuzzy inference system.

2.6.1 Rule Base

The rules base, sometimes called a *linguistic model,* is a set of fuzzy rules $R^{(k)}$, $k = 1,...,N$, of the form

$$\begin{aligned} R^{(k)} : \ & \textbf{IF } x_1 \text{ is } A_1^k \textbf{ AND } x_2 \text{ is } A_2^k \textbf{ AND...AND} \\ & x_n \text{ is } A_n^k \textbf{ THEN } y_1 \text{ is } B_1^k \textbf{ AND } y_2 \text{ is } B_2^k \textbf{ AND...AND } y_m \text{ is } B_m^k, \end{aligned} \tag{2.91}$$

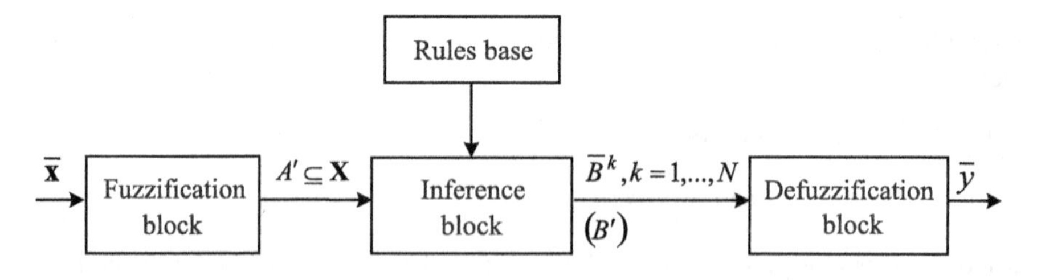

Fig. 2.3 Fuzzy inference system [10]

where N is the number of fuzzy rules, A_i^k – fuzzy sets such as

$$A_i^k \subseteq \mathbf{X}_i \subset \mathbf{R}, \quad i = 1, ..., n, \tag{2.92}$$

B_j^k – fuzzy sets such as

$$B_j^k \subseteq \mathbf{Y}_j \subset \mathbf{R}, \quad j = 1, ..., m, \tag{2.93}$$

$x_1, x_2, ..., x_n$ – input variables of the linguistic model, while

$$[x_1, x_2, ..., x_n]^T = \mathbf{x} \in \mathbf{X}_1 \times \mathbf{X}_2 \times \ldots \times \mathbf{X}_n, \tag{2.94}$$

$y_1, y_2, ..., y_m$ – output variables of the linguistic model, while

$$[y_1, y_2, ..., y_m]^T = \mathbf{y} \in \mathbf{Y}_1 \times \mathbf{Y}_2 \times \ldots \times \mathbf{Y}_m. \tag{2.95}$$

Symbols $\mathbf{X}_i$, $i = 1, ..., n$, and $\mathbf{Y}_j$, $j = 1, ..., m$, denote the spaces of input and output variables, respectively.

Further in our discussion, we shall assume that particular rules $R^{(k)}$, $k = 1, ..., N$, are related to each other using a logical operator "or". Moreover, we shall assume that outputs $y_1, y_2, ..., y_m$ are mutually independent. Therefore, without losing the generality, we shall consider fuzzy rules with a scalar output of the form

$$\begin{aligned} R^{(k)}: \textbf{IF } x_1 \text{ is } A_1^k \textbf{ AND } x_2 \text{ is } A_2^k \textbf{ AND} \\ \ldots \textbf{AND } x_n \text{ is } A_n^k \textbf{ THEN } y \text{ is } B^k, \end{aligned} \tag{2.96}$$

where $B^k \subseteq \mathbf{Y} \subset \mathbf{R}$ and $k = 1, ..., N$. Let us notice that any rule of the above form) consists of the part **IF**, called *antecedent*, and a part **THEN** called *consequent*. The antecedent of the rule contains a set of conditions, and the consequent contains the inference. Variables $\mathbf{x} = [x_1, x_2, ..., x_n]^T$ and y may take both linguistic values defined in words, like "small", "average" and "high", and numerical values. Let us denote

$$\mathbf{X} = \mathbf{X}_1 \times \mathbf{X}_2 \times \ldots \times \mathbf{X}_n \tag{2.97}$$

and

$$A^k = A_1^k \times A_2^k \times \ldots \times A_n^k. \tag{2.98}$$

Applying the above notations, rule (2.96) may be presented in the form

$$R^{(k)} : A^k \to B^k, \quad k = 1, \ldots, N. \tag{2.99}$$

Let us notice that the rule $R^{(k)}$ may be interpreted as a fuzzy relation defined on the set $\mathbf{X} \times \mathbf{Y}$ i.e. $R^{(k)} \subseteq \mathbf{X} \times \mathbf{Y}$ is a fuzzy set with membership function

$$\mu_{R^{(k)}}(\mathbf{x}, y) = \mu_{A^k \to B^k}(\mathbf{x}, y). \tag{2.100}$$

When designing fuzzy controllers, it should be decided whether the number of rules is sufficient, whether they are consistent and whether there are interactions between particular rules.

2.6.2 Fuzzification Block

A control system with fuzzy logic operates on fuzzy sets. That is why a specific value $\overline{\mathbf{x}} = [\overline{x}_1, \overline{x}_2, ..., \overline{x}_n]^T \in \mathbf{X}$ of an input signal of the fuzzy controller is subject to a fuzzification operation, as a result of which it is mapped into a fuzzy set $A' \subseteq \mathbf{X} = \mathbf{X}_1 \times \mathbf{X}_2 \times ... \times \mathbf{X}_n$. Most often in case of control problems, singleton type fuzzification is applied as follows

$$\mu_{A'}(\mathbf{x}) = \delta(\mathbf{x} - \overline{\mathbf{x}}) = \begin{cases} 1, \text{ if } \mathbf{x} = \overline{\mathbf{x}}, \\ 0, \text{ if } \mathbf{x} \neq \overline{\mathbf{x}}. \end{cases} \tag{2.101}$$

The fuzzy set A' is an input of the inference block. If the input signal is measured together with the interference (noise), then the fuzzy set A' may be defined using the membership function

$$\mu_{A'}(\mathbf{x}) = \exp\left[-\frac{(\mathbf{x} - \overline{\mathbf{x}})^T (\mathbf{x} - \overline{\mathbf{x}})}{\delta}\right], \tag{2.102}$$

where $\delta > 0$.

2.6.3 Inference Block

Let us assume that at the input to the inference block we have a fuzzy set $A' \subseteq \mathbf{X} = \mathbf{X}_1 \times \mathbf{X}_2 \times ... \times \mathbf{X}_n$. We shall find an appropriate fuzzy set at the output of this block. Let us consider two cases to which different defuzzification methods will correspond.

Case 1. At the output of the inference block we obtain N fuzzy sets $\overline{B}^k \subseteq \mathbf{Y}$ according to the generalized fuzzy *modus ponens* inference rule. The fuzzy set $\overline{B}^k$ is determined by the composition of the fuzzy set A' and the relation $R^{(k)}$, i.e.

$$\overline{B}^{(k)} = A' \circ \left(A^k \to B^k\right), \quad k = 1, \ldots, N. \tag{2.103}$$

Using Definition 2.17, we shall determine the membership function of the fuzzy set $\overline{B}^k$ as follows

$$\mu_{\overline{B}^k}(y) = \sup_{\mathbf{x} \in \mathbf{X}} \left[\mu_{A'}(\mathbf{x}) \overset{T}{*} \mu_{A^k \to B^k}(\mathbf{x}, y)\right]. \tag{2.104}$$

A specific form of the function $\mu_{\overline{B}^k}(y)$ depends on the chosen t-norm, inference rule and on the method of defining the Cartesian product of fuzzy sets. We should note that in case of singleton type fuzzification, formula (2.104) takes the form

$$\mu_{\overline{B}^k}(y) = \mu_{A^k \to B^k}(\overline{\mathbf{x}}, y). \tag{2.105}$$

Case 2. At the output of the inference block, we obtain one fuzzy set $B' \subseteq \mathbf{Y}$, defined by the formula

$$B' = \bigcup_{k=1}^{N} A' \circ R^{(k)} = \bigcup_{k=1}^{N} A' \circ \left(A^k \to B^k\right). \tag{2.106}$$

We obtain the membership function of a fuzzy set B'

$$\mu_{B'}(y) = \overset{N}{\underset{k=1}{S}} \mu_{\overline{B}^k}(y), \tag{2.107}$$

while the membership function $\mu_{\overline{B}^k}(y)$ is given by formula (2.104).

2.6.4 Defuzzification Block

The output value of the inference block is either N fuzzy sets $\overline{B}^k$ with membership functions $\mu_{\overline{B}^k}(y)$, $k = 1, 2, \ldots, N$, or a single fuzzy set B' with membership function $\mu_{B'}(y)$. Now we consider a problem of mapping fuzzy sets $\overline{B}^k$ (or fuzzy set B') into a single value $\overline{y} \in \mathbf{Y}$. This mapping is called d*efuzzification* and it is made in the defuzzification block.

If the output value of the inference block is N fuzzy sets $\overline{B}^k$, then the value $\overline{y} \in \mathbf{Y}$ may be determined using the following methods:

1. *Center average defuzzification method*. The value $\overline{y}$ shall be calculated using the formula

$$\overline{y} = \frac{\sum_{k=1}^{N} \mu_{\overline{B}^k}(\overline{y}^k)\, \overline{y}^k}{\sum_{k=1}^{N} \mu_{\overline{B}^k}(\overline{y}^k)}, \tag{2.108}$$

where $\overline{y}^k$ is the point in which the function $\mu_{B^k}(y)$ takes the maximum value, i.e.

$$\mu_{B^k}\left(\overline{y}^k\right) = \max_{y} \mu_{B^k}(y). \tag{2.109}$$

Point $\overline{y}^k$ is called *center* of the fuzzy set B^k. Let us notice that value $\overline{y}$ does not depend on the shape and support of the membership function $\mu_{B^k}(y)$.
2. *Center of sums defuzzification method*. The value $\overline{y}$ is computed as follows:

$$\overline{y} = \frac{\int_{\mathbf{Y}} y \sum_{k=1}^{N} \mu_{\overline{B}^k}(y)\, \mathrm{d}y}{\int_{\mathbf{Y}} \sum_{k=1}^{N} \mu_{\overline{B}^k}(y)\, \mathrm{d}y}. \tag{2.110}$$

If the output value of the inference block is a single fuzzy set B', then the value $\overline{y}$ may be determined using the following methods:

3. *Center of gravity method* (or *center of area method*). The value $\overline{y}$ is calculated as *the center of gravity* of the membership function $\mu_{B'}(y)$, i.e.

$$\overline{y} = \frac{\int_{\mathbf{Y}} y\mu_{B'}(y)\,\mathrm{d}y}{\int_{\mathbf{Y}} \mu_{B'}(y)\,\mathrm{d}y} = \frac{\int_{\mathbf{Y}} y\, S_{k=1}^{N}\, \mu_{\overline{B}^k}(y)}{\int_{\mathbf{Y}} S_{k=1}^{N}\, \mu_{\overline{B}^k}(y)}, \tag{2.111}$$

assuming that both integrals in the above formula exist. In a discrete case, the above formula takes the form

$$\overline{y} = \frac{\sum_{k=1}^{N} \mu_{B'}(\overline{y}^k)\, \overline{y}^k}{\sum_{k=1}^{N} \mu_{B'}(\overline{y}^k)}. \tag{2.112}$$

4. *Maximum membership function method.* The value $\overline{y}$ is computed according to the formula

$$\mu_{B'}(\overline{y}) = \sup_{y\in\mathbf{Y}} \mu_{B'}(y) \tag{2.113}$$

assuming that $\mu_{B'}(y)$ is a unimodal function. This method does not consider the shape of the membership function.

2.7 Fuzzy Classification

One of soft computing applications is classification which consists in assigning an object described by a set of features to a class. The object $x \in \mathbf{X}$ is described by the vector of features $\mathbf{v} \in \mathbf{V}$. Thus we can equate object x class membership with its feature values $\overline{\mathbf{v}} = [\overline{v}_1, \overline{v}_2, \ldots, \overline{v}_n]$. Consequently, we can use interchangeably x or $\overline{\mathbf{v}}$. Let us assume that fuzzy set $A \subseteq \mathbf{V}$ is given as its membership function $\mu_A(x) = \mu_A(\overline{\mathbf{v}}) = \mu_A(\overline{v}_1, \overline{v}_2, \ldots, \overline{v}_n)$ where $\overline{v}_i \in \mathbf{V}_i$ for $i = 1, \ldots, n$. We also define the set of all object x features $Q = \{v_1, v_2, \ldots, v_n\}$. Fuzzy classifiers are frequently used thanks to their ability to use knowledge in the form of intelligible IF-THEN fuzzy rules. The standard form of fuzzy rules are suitable for approximation and majority of control tasks. In the case of classification tasks, rules in other form are more appropriate, i.e.

$$\begin{array}{c} R^r: \text{IF } v_1 \text{ is } A_1^r \text{ AND } v_2 \text{ is } A_2^r \text{ AND} \ldots \\ \ldots \text{ AND } v_n \text{ is } A_n^r \\ \text{THEN } x \in \omega_1(\overline{z}_1^r), x \in \omega_2(\overline{z}_2^r), \ldots, x \in \omega_m(\overline{z}_m^r) \end{array}, \tag{2.114}$$

where $r = 1, \ldots, N$, N is the number of rules and $\overline{z}_j^r$ is the membership degree of the object x to the j–th class ω_j according to rule r. Let us assume that the membership of objects to classes is not fuzzy but crisp, i.e.

$$\overline{z}_j^r = \begin{cases} 1 & \text{if } x \in \omega_j \\ -1 & \text{if } x \notin \omega_j \end{cases}. \tag{2.115}$$

We write just $x \in \omega_j$ when $\overline{z}_j^r = 1$ (which means that object x belongs to the j–th class, according to the r–th rule) in definition of the r–th rule. We can omit the part $x \in \omega_j(\overline{z}_j^r)$ when $\overline{z}_j^r = -1$ (what means that object x does not belong to the j–th class, according to the r–th rule). In the case when values of all features v_i of classified object x are available, it is easy to adopt any formula of neuro-fuzzy system to classification tasks.

2.8 Fuzzy c-Means

Now let us present the fuzzy c-means algorithm which allows assigning the same objects to various clusters with appropriate membership degrees. The FCM algorithm is the most frequently used algorithm of fuzzy clustering. It detects clusters with prototypes which are points in the data space. All clusters have the same shape dependent on the norm chosen in advance since the algorithm has no possibility to adjust the matrix $\mathbf{A}$ to existing data. This algorithm is derived by minimization of the criterion

$$J(\mathbf{X};\mathbf{U},\mathbf{V}) = \sum_{i=1}^{c} \sum_{k=1}^{M} (\mu_{ik})^m \|\mathbf{x}_k - \mathbf{v}_i\|_{\mathbf{A}}^2, \tag{2.116}$$

where

$$\mathbf{U} = [\mu_{ik}] \in Z_2 \tag{2.117}$$

is the matrix of the set $\mathbf{X}$ partition, whereas

$$\mathbf{V} = [\mathbf{v}_1, \mathbf{v}_2, \ldots, \mathbf{v}_c] \tag{2.118}$$

is the vector of centers which are to be defined as a result of the algorithm operation, $\mathbf{v}_i \in R^n$, $i = 1,...,c$. The following term appearing in formula (2.116)

$$D_{ik\mathbf{A}}^2 = \|\mathbf{x}_k - \mathbf{v}_i\|_{\mathbf{A}}^2 = (\mathbf{x}_k - \mathbf{v}_i)^T \mathbf{A} (\mathbf{x}_k - \mathbf{v}_i) \tag{2.119}$$

permits to compute the distance between vector $\mathbf{x}_k$ and cluster center $\mathbf{v}_i$, and $m \in (1, \infty)$ is a coefficient indicating the fuzziness degree of formed clusters. When $m \to 1$, the partition becomes less and less fuzzy. When $m \to \infty$, the partition becomes more and more fuzzy (then $\mu_{ik} = 1/c$). In practice the value $m = 2$ is often chosen. In order to execute the algorithm, having a given data set $\mathbf{X}$, we must choose the number of clusters c, fuzziness degree m, parameter ε in the algorithm stopping criterion and initiate at random matrix $\mathbf{U}^{(0)} \in Z_2$ and vector of clusters prototypes $\mathbf{V}^{(0)}$. The algorithm stopping criterion is the same as in case of the Hard c-Means algorithm (see [6]). The FCM algorithm, like Hard c-Means, may give various results depending on the initialization. The shape of clusters depends on the adopted distance measure.

2.9 Summary and Discussion

This chapter presented basic definitions of fuzzy set theory as well as design of fuzzy systems. Fuzzy systems can be better suited to certain tasks than other soft computing methods. Initial set of rules can be picked by an expert or by machine learning methods. For example we can use clustering algorithms to set fuzzy linguistic values on the basis of numerical data. The chapter presented also alternative form of fuzzy rules suitable for classification.

References

1. Bezdek, J., Keller, J., Krisnapuram, R., Pal, N.: Fuzzy Models and Algorithms for Pattern Recognition and Image Processing. Kluwer Academic Press (1999)
2. Bezdek, J.C.: Fuzzy Models for Pattern Recognition. IEEE Press, New York (1992)
3. Czogała, E., Łęski, J.: Fuzzy and Neuro-Fuzzy Intelligent Systems. Springer, New York (2000)
4. Dubois, D., Prade, H.: Fuzzy sets and systems - Theory and applications. Academic press, New York (1980)
5. Jang, R.J.S., Sun, C.T., Mizutani, E.: Neuro-Fuzzy and Soft Computing, A Computational Approach to Learning and Machine Intelligence. Prentice-Hall, Upper Saddle River (1997)
6. Kuncheva, L.: Combining Pattern Classifiers. STUDFUZZ. John Wiley & Sons (2004)
7. Nauck, D.: Foundations of Neuro-Fuzzy Systems. John Wiley, Chichester (1997)
8. Pedrycz, W.: Fuzzy Control and Fuzzy Systems. Research Studies Press, London (1989)
9. Rutkowski, L.: Flexible Neuro-Fuzzy Systems. Kluwer Academic Publishers (2004)
10. Rutkowski, L.: Computational Intelligence Methods and Techniques. Springer, Heidelberg (2008)
11. Zadeh, L.A.: Fuzzy sets. Information Control 8, 338–353 (1965)
12. Zadeh, L.A.: The concept of a linguistic variable and its application to approximate reasoning. Inf. Sci. 8(3), 199–249 (1975)

Chapter 3
Ensemble Techniques

Combining single classifiers into larger ensembles is an established method for improving the accuracy. Of course, the obvious improvement is bound up with increased storage space (memory) and computational burden. However this trade-off is easy to accept with modern computers. Classifiers can be combined at the level of features or data subsets and by the use of different classifiers or different combiners, see Figure 3.1. Popular methods are bagging and boosting which are meta algorithms for learning different classifiers.

3.1 Bagging

Bagging [8, 13] is a procedure for combining classifiers generated using the same training set. The name comes from Bootstrap AGGregatING. The ensemble members use data sets created from the original data set by drawing at random with replacement. There are more sophisticated metalearning procedures for ensembles such as the AdaBoost, described in the next section. They allow to achieve better accuracy but bagging can be used without any modifications in parallel computing as all classifiers work independently and only during classification stage the overall response (decision) is computed. Moreover the algorithm is suitable for multiclass problems. Bagging produces replicates of the training set $\mathbf{z}$ and trains a classifier D_k on each replicate S_k. Each classifier is applied to a test pattern $\mathbf{x}$ which is classified on a majority vote basis, ties being resolved arbitrarily. We have a set of labels $\Omega = \{\omega_1, \omega_2, \ldots, \omega_C\}$, where C is the number of possible classes, labeled ω_i, $i = 1, \ldots, C$. We consider the ensemble of classifiers $\mathbf{D} = [D_1, \ldots, D_J]$, where there are J base classifiers D_k, $k = 1, \ldots, J$. We assume that the output of classifier D_k is $\mathbf{d}_k(\mathbf{x}) = [d_{k,1}(\mathbf{x}), \ldots, d_{k,C}(\mathbf{x})]^T \in \{0,1\}^C$, where $d_{k,j} = 1$ if D_k determines that $\mathbf{x}$ belong to class ω_j, and $d_{k,j} = 0$ otherwise. The majority vote will result in an ensemble decision for class ω_k if

$$\sum_{i=1}^{J} d_{i,k}(\mathbf{x}) = \max_{1 \leq j \leq C} \sum_{i=1}^{J} d_{i,j}(\mathbf{x}) \tag{3.1}$$

R. Scherer: Multiple Fuzzy Classification Systems, STUDFUZZ 288, pp. 29–37.
springerlink.com

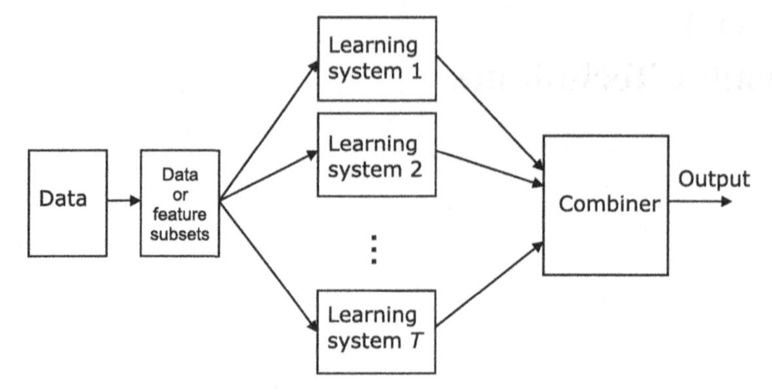

Fig. 3.1 General diagram of various techniques to combine classifiers

The Bagging algorithm consists of the following steps

1. Initialize the parameters
   - the ensemble $\mathbf{D} = \emptyset$
   - the number of classifiers to train J
2. For $k = 1, \dots, J$ repeat points 3-5
3. Take sample S_k from original dataset $\mathbf{Z}$
4. Build a classifier D_k using S_k as the training set
5. Add the classifier to the current ensemble $\mathbf{D} = \mathbf{D} \cup D_k$
6. Return $\mathbf{D}$ as algorithm outcome
7. Run $\mathbf{x}$ on the input $D_1, \dots, D_J$
8. The vector $\mathbf{x}$ is a member of class ω_k, if condition (3.1) is fulfilled.

3.2 Boosting

This section describes the AdaBoost algorithm which is the most popular boosting method [3, 11, 12]. The algorithm described here is designed for binary classification. Let us denote the l-th learning vector by $\mathbf{z}^l = [x_1^l, ..., x_n^l, y^l]$, $l = 1...m$ is the number of a vector in the learning sequence, n is the dimension of input vector $\mathbf{x}^l$, and y^l is the learning class label. Weights D^l assigned to learning vectors, have to fulfill the following conditions

$$\begin{aligned} &\text{(i)}\ 0 < D^l < 1\ , \\ &\text{(ii)}\ \sum_{l=1}^{m} D^l = 1\ . \end{aligned} \tag{3.2}$$

The weight D^l is the information how well classifiers were learned in consecutive steps of an algorithm for a given input vector x^l. Vector $\mathbf{D}$ for all input vectors is initialized according to the following equation

$$D_t^l = \frac{1}{m}, \quad \text{for } t = 0,...,T \,, \tag{3.3}$$

where t is the number of a boosting iteration (and a number of a classifier in the ensemble). Let $\{h_t(\mathbf{x}) : t = 1,...,T\}$ denotes a set of hypotheses obtained in consecutive steps t of the algorithm being described. For simplicity we limit our problem to a binary classification (dichotomy) i.e. $y \in \{-1, 1\}$ or $h_t(\mathbf{x}) = \pm 1$. Similarly to learning vectors weights, we assign a weight c_t for every hypothesis, such that

$$\begin{array}{l} \text{(i)} \sum_{t=1}^{T} c_t = 1 \,, \\ \text{(ii)}\ c_t > 0 \,. \end{array} \tag{3.4}$$

Now in the AdaBoost algorithm we repeat steps 1-4 for $t = 1, \ldots, T$:

1. Create hypothesis h_t and train it with a data set with respect to a distribution d_t for input vectors.
2. Compute the classification error ε_t of a trained classifier h_t according to the formula

$$\varepsilon_t = \sum_{l=1}^{m} D_t^l(z^l) I(h_t(\mathbf{x}^l) \neq y^l) \,, \tag{3.5}$$

where I is the indicator function

$$I(a \neq b) = \begin{cases} 1 \text{ if } a \neq b \\ 0 \text{ if } a = b \end{cases} . \tag{3.6}$$

If $\varepsilon_t = 0$ or $\varepsilon_t \geq 0.5$, stop the algorithm.
3. Compute the value

$$\alpha_t = 0.5 \ln \frac{1 - \varepsilon_t}{\varepsilon_t} \,. \tag{3.7}$$

4. Modify weights for learning vectors according to the formula

$$D_{t+1}(\mathbf{z}^l) = \frac{D_t(\mathbf{z}^l) \exp\{-\alpha_t \mathbf{I}(h_t(\mathbf{x}_l) = y^l)\}}{N_t} \,, \tag{3.8}$$

where N_t is a constant such that $\sum_{l=1}^{m} D_{t+1}(\mathbf{z}^l) = 1$. To compute the overall output of the ensemble of classifiers trained by AdaBoost algorithm, the following formula is used

$$f(\mathbf{x}) = \sum_{t=1}^{T} c_t h_t(\mathbf{x}) \,, \tag{3.9}$$

where

$$c_t = \frac{\alpha_t}{\sum_{t=1}^{T} \alpha_t} \tag{3.10}$$

is classifier importance for a given training set, $h_t(\mathbf{x})$ is the response of the hypothesis t on the basis of feature vector $\mathbf{x} = [x_1, ..., x_n]$. The coefficient c_t value is computed on the basis of the classifier error and can be interpreted as the measure of classification accuracy of the given classifier. Moreover, the assumption (3.2) should be met. As we see, the AdaBoost algorithm is a meta-learning algorithm and does not determine the way of learning for classifiers in the ensemble.

3.3 Negative Correlation Learning

Negative correlation learning [9, 10] is a meta learning algorithm for creating negatively correlated ensembles. Let us denote the l-th learning vector by $\mathbf{z}^l = [x_1^l, ..., x_n^l, y^l]$, $l = 1...m$ is the number of a vector in the learning sequence, n is the dimension of input vector $\mathbf{x}^l$, and y^l is the learning class label. The overall output of the ensemble of classifiers is computed by averaging outputs of all hypothesis

$$f(\mathbf{x}) = \frac{1}{T}\sum_{t=1}^{T} h_t(\mathbf{x}) , \tag{3.11}$$

where $h_t(\mathbf{x})$ is the response of the hypothesis t on the basis of feature vector $\mathbf{x} = [x_1, ..., x_n]$. All neuro-fuzzy parameters, i.e. antecedent and consequent fuzzy sets parameters, are tuned by the backpropagation algorithm [14]. Having given learning data set of pair $(\mathbf{x}^l, y^l)$, where y^l is the desired response of the system we can use the following error measure

$$E\left(\mathbf{x}^l, y^l\right) = \frac{1}{2}\left[h_t(\mathbf{x}) - y^l\right]^2 . \tag{3.12}$$

Every neuro-fuzzy system parameter, denoted for simplicity as w, can be determined by minimizing the error measure in the iterative procedure. For every iteration t, the parameter value is computed by

$$w(t+1) = w(t) - \eta \frac{\partial E\left(\mathbf{x}^l, y^l; t\right)}{\partial w(t)} , \tag{3.13}$$

where η is a learning coefficient. This is a standard gradient learning procedure. As we build an ensemble of negatively correlated neuro-fuzzy systems, the error measure is modified by introducing a penalty term $p_t(l)$ and determining error after the whole epoch

$$E_t = \frac{1}{m}\sum_{l=1}^{m} E_t(l) = \frac{1}{m}\sum_{l=1}^{m}\frac{1}{2}\left(h_t(l) - y^l\right)^2 + \frac{1}{m}\sum_{l=1}^{m}\lambda p_t(l) , \tag{3.14}$$

where λ is a coefficient responsible for the strength of decorrelation. The penalty term is defined

$$p_t(l) = (h_t(l) - f(\mathbf{x}))\sum_{k \neq l}(h_t(k) - f(\mathbf{x})) . \tag{3.15}$$

The NCL metalearning tries to keep responses of the member neuro-fuzzy systems as different as possible, retaining at the same time classification accuracy.

3.4 Backpropagation Learning in Boosting Neuro-fuzzy Ensembles

In this section we show a modification of the backpropagation algorithm for speeding up learning in boosting ensembles. In the first step of the AdaBoost algorithm there exists imprecise statement "train it with a data set with respect to the distribution for input vectors **x**". There exists several methods for taking into account weights D_t^q, $t = 1,\ldots,T$, $q = 1,\ldots,m$ in the process of learning. Please note that $q = l$ for the first epoch. There are several methods to take into account the input vector weights during training. Let us assume that our classifiers are neural networks or neuro-fuzzy systems trained by the backpropagation algorithm. One of the methods is based on picking samples at random with the probability D_t^q being equal to weights of samples $\mathbf{x}^q$, $q = 1,\ldots,m$, and in this way creating a new learning set for the hypothesis h_t. Unfortunately this method is nondeterministic and researchers obtain different results using it. In [3] it is recommended to use, for each hypothesis, modified error function which depends on weight values for learning vectors. The basis for further work can be a formula given in [11], where standard error in the backpropagation algorithm is multiplied by sample weights

$$Q^q(l) = (y^q(l) - h_t^q(\mathbf{x}^q(l)))^2 D_t^q(l),\ q = 1,...,M,\ t = 1,...,T. \tag{3.16}$$

This approach has the following drawback. Let us assume that in a learning set there is a large number of vectors, say 10 000. From (3.16) it results that weight values in initial steps of AdaBoost will be very small, e.g. $d_t^q = \frac{1}{10000}$ for $t = 1$. Error values used during training are very small, so the training will be quite long. Please note that the weights are modified after a given classifier is trained. To overcome the problem a new method for incorporating sample weights will be proposed [7, 6]. The method will be presented using the Mamdani neuro-fuzzy systems which are described in detail in Chapter 5. Output of such systems is given by

$$y = \frac{\sum\limits_{r=1}^{N} \bar{y}^r \cdot \mu_{\bar{B}^r}(\bar{y}^r)}{\sum\limits_{r=1}^{N} \mu_{\bar{B}^r}(\bar{y}^r)}, \tag{3.17}$$

where membership functions of fuzzy sets $\bar{B}^r$, $r = 1,\ldots,N$, are given by

$$\mu_{\bar{B}^r}(y) = \sup_{\mathbf{x}\in\mathbf{X}} \left[\mu_{A'}(x) \overset{T}{*} \mu_{A^r \to B^r}(x,y) \right]. \tag{3.18}$$

If we assume that we use singleton defuzzification then above formula takes the following form

$$\mu_{\bar{B}^r}(y) = \mu_{A^r \to B^r}(\bar{x},y) = T\left(\mu_{A^r}(\bar{x}), \mu_{B^r}(y)\right). \tag{3.19}$$

Let us observe that

$$\mu_{A^r}(\bar{x}) = \overset{n}{\underset{i=1}{T}} \left(\mu_{A_i^r}(\bar{x}_i) \right) , \tag{3.20}$$

and assume that

$$\mu_{B^r}(\bar{y}^r) = 1 . \tag{3.21}$$

Then, using a t-norm property ($T(a,1) = a$), we obtain a formula describing a single Mamdani fuzzy system

$$\bar{y} = \frac{\sum_{r=1}^{N} \bar{y}^r \cdot \overset{n}{\underset{i=1}{T}} \left(\mu_{A_i^r}(\bar{x}_i) \right)}{\sum_{r=1}^{N} \overset{n}{\underset{i=1}{T}} \left(\mu_{A_i^r}(\bar{x}_i) \right)} . \tag{3.22}$$

Linguistic variables are described by the Gaussian membership function

$$\mu_{A_i^r}(x_i) = \exp\left[-\left(\frac{x_i - \bar{x}_i^r}{\sigma_i^r} \right)^2 \right] . \tag{3.23}$$

If we choose product t-norm, then the neuro-fuzzy system with the center average defuzzification and Mamdani-type relation is described by

$$\bar{y} = \frac{\sum_{r=1}^{N} \bar{y}^r \left[\prod_{i=1}^{n} \exp\left(-\left(\frac{x_i - \bar{x}_i^r}{\sigma_i^r} \right)^2 \right) \right]}{\sum_{r=1}^{N} \left[\prod_{i=1}^{n} \exp\left(-\left(\frac{x_i - \bar{x}_i^r}{\sigma_i^r} \right)^2 \right) \right]} . \tag{3.24}$$

Let us denote a pair of input vector and output signal by $\mathbf{x}^q$ and d^q, respectively, where q denotes vector number, $q = 1, \ldots, M$. Let us define the error

$$Q^q = \frac{1}{2} \left[f(\mathbf{x}^q) - d^q \right]^2 . \tag{3.25}$$

During learning we tune parameters $\bar{y}^r, \bar{x}_i^r, \sigma_i^r$ to minimize (3.25) for all learning vectors. The most common approach is the use of gradient algorithms. Value of $\bar{y}^r$ parameter is computed by

$$\bar{y}^r(l+1) = \bar{y}^r(l) - \eta \frac{\partial Q}{\partial \bar{y}^r} |l , \tag{3.26}$$

where $l = 0, 1, 2...$ is a number of step of gradient algorithm, and η is a learning coefficient. Output signal value depends on $\bar{y}^r$ only by the numerator of (3.24). Let us assume that $a = \sum_{r=1}^{N} \bar{y}^r z^r$, $b = \sum_{r=1}^{N} z^r$ and $z^r = \prod_{i=1}^{n} \exp\left(-\left(\frac{x_i - \bar{x}_i^j}{\sigma_i^j} \right)^2 \right)$. Then we can write modification of the parameter $\bar{y}^r$ for consecutive steps as follows

$$\bar{y}^r(l+1) = \bar{y}^r(l) - \eta \frac{f-d}{b} z^r . \tag{3.27}$$

An analogical deliberation for parameters $\bar{x}_i^r$ and σ_i^r yields

$$\bar{x}_i^r(l+1) = \bar{x}_i^r(l) - \eta\frac{f-d}{b}(\bar{y}^r - f)z^r\frac{2(x_i^q - \bar{x}_i^r(l))}{\sigma_i^{r2}(l)}, \tag{3.28}$$

$$\sigma_i^r(l+1) = \sigma_i^r(l) - \eta\frac{f-d}{b}(\bar{y}^r - f)z^r\frac{2(x_i^q - \bar{x}_i^r(l))^2}{\sigma_i^{r3}(l)}. \tag{3.29}$$

Formulas (3.27-3.29) can be treated as a special case of the backpropagation algorithm. If we would like to build a modular learning system based on boosting consisted of several neuro-fuzzy systems with gradient learning, we end up with similar problems as in modular boosting neural networks. Learning coefficient for a sample q is computed in every step t of the algorithm. We tend to a situation where samples with maximal weight values have the same learning coefficient as set by the researcher, i.e. $v_t^q = \eta$ for samples for which $D_t^{\max} = \max_{i=1,\ldots,M} D_t^i$ where m is the number of learning vectors. For the rest of samples the learning coefficient should be respectively smaller. To fulfill the above assumptions we propose the following modification of the learning coefficient for all learning vectors

$$v_t^q = \eta\frac{D_t^q + 1}{D_t^{\max} + 1}, \tag{3.30}$$

where $D_t^{\max}$ is the maximal value of sample weights for t-th hypothesis. Finally we obtain the new formulas for parameters update in an ensemble of the Mamdani neuro-fuzzy systems

$$\begin{aligned} \bar{y}_t^r(l+1) = \bar{y}_t^r(l) - v_t^q \tfrac{f_t-d}{b_t} z_t^r = \\ \bar{y}_t^r(l) - \left(\eta \tfrac{D_t^q+1}{D_t^{\max}+1}\right) \tfrac{f_t-d}{b_t} z_r^t, \end{aligned} \tag{3.31}$$

$$\begin{aligned} \bar{x}_i^{(r)t}(l+1) = \bar{x}_i^{(r)t}(l) - v_t^q \tfrac{f_t-d}{b_t}(\bar{y}_t^r - f_t)z_t^r \tfrac{2(x_i^{(p)t} - \bar{x}_i^{(r)t}(l))}{\sigma_i^{(r)t2}(l)} = \\ \bar{x}_i^{(r)t}(l) - \left(\eta \tfrac{D_t^q+1}{D_t^{\max}+1}\right) \tfrac{f_t-d}{b_t}(\bar{y}_t^r - f_t)z_t^r \tfrac{2(x_i^{(p)t} - \bar{x}_i^{(r)t}(l))}{\sigma_i^{(r)t2}(l)}, \end{aligned} \tag{3.32}$$

$$\begin{aligned} \sigma_i^{(r)t}(l+1) = \sigma_i^{(r)t}(l) - v_t^q \tfrac{f_t-d}{b_t}(\bar{y}^r - f_t)z_t^r \tfrac{2(x_i^{(p)t} - \bar{x}_i^{(r)t}(l))^2}{\sigma_i^{(r)t3}(l)} = \\ \sigma_i^{(r)t}(l) - \left(\eta \tfrac{D_t^q+1}{D_t^{\max}+1}\right) \tfrac{f_t-d}{b_t}(\bar{y}_t^r - f_t)z_t^r \tfrac{2(x_i^{(p)t} - \bar{x}_i^{(r)t}(l))^2}{\sigma_i^{(r)t3}(l)}, \end{aligned} \tag{3.33}$$

where d is desired output and f is the output of a neuro-fuzzy system. In the next section we present a modification of the fuzzy c-means clustering algorithm for boosting ensemble modification.

3.5 Fuzzy c-Means Modification for Boosting Initialization

One of several advantages of fuzzy systems is the possibility of interpretation of the knowledge. Thanks to this we can also initialize the parameters of the fuzzy system. A common method for such initialization is the use of the fuzzy c-means algorithm [1, 2, 4]. The FCM algorithm is based on fuzzy clustering of data vectors and assigning membership values to them. The algorithm is derived by minimizing the criterion [5]

$$J(\mathbf{X};\mathbf{U};\mathbf{V}) = \sum_{i=1}^{C}\sum_{q=1}^{M} (\mu_{iq})^m \left\|\mathbf{x}_q - \mathbf{v}_i\right\|_{\mathbf{A}}^2 \tag{3.34}$$

and $\mathbf{U} = [\mu_{iq}] \in Z$ is a partition matrix of dataset $\mathbf{X}$, and $\mathbf{V} = [\mathbf{v}_1, \mathbf{v}_2, ..., \mathbf{v}_C]$ is a vector of centers to be determined by the algorithm, $\mathbf{v}_i \in R^n$, $i = 1, \dots, C$. The algorithm can be stopped when matrix $\mathbf{U}$ does not change or the change is below certain level $\left(\left\|\mathbf{U}^{(l+1)} - \mathbf{U}^{(l)}\right\| < \varepsilon\right)$. Alternatively we can check the change of centers, i.e. $\left\|\mathbf{V}^{(l+1)} - \mathbf{V}^{(l)}\right\| < \varepsilon$, where l the iteration number in the FCM algorithm. We modify the FCM algorithm by adding sample weight D_t^q to the formula for computing data objects memberships, see Fig. 3.2. This causes to take into account boosting sample weights during initial setting linguistic values (fuzzy sets) by clustering.

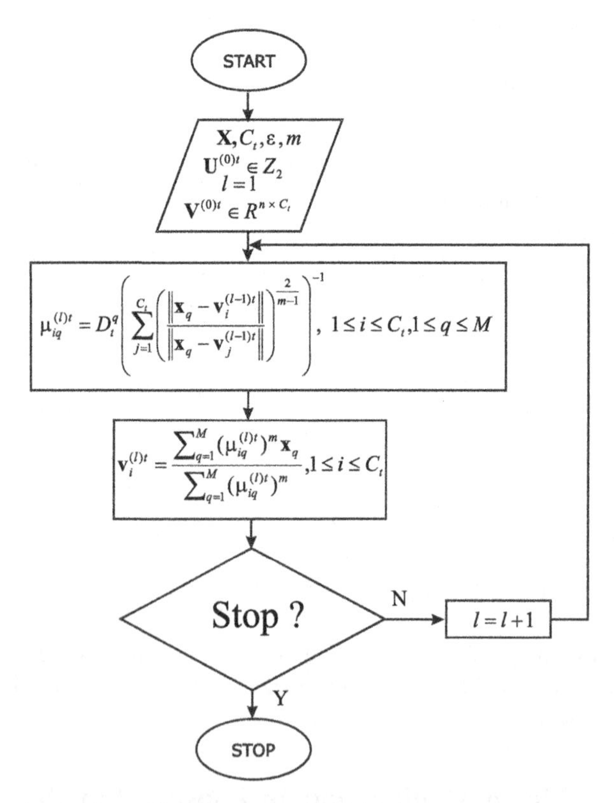

Fig. 3.2 Flowchart of the modified fuzzy c-means algorithm

3.6 Summary and Discussion

This chapter described some popular ensemble methods. All of them are metalearning methods and they have to use some other "standard" algorithms to learn members of the ensemble. In this book, the hypothesis within the ensemble are trained by the fuzzy c-means algorithm and the backpropagation algorithm. Thanks to inherent interpretability of fuzzy systems, it is easy to initialize the system using some expert knowledge or exploratory data analysis methods.

References

1. Bezdek, J., Keller, J., Krisnapuram, R., Pal, N.: Fuzzy Models and Algorithms for Pattern Recognition and Image Processing. Kluwer Academic Press (1999)
2. Bezdek, J.C., Pal, S.: Fuzzy Models for Pattern Recognition. IEEE Press, New York (1992)
3. Breiman, L.: Bias, variance, and arcing classifiers. Tech. Rep. Technical Report 460, Statistics Department, University of California (1997)
4. Czogała, E., Łęski, J.: Fuzzy and Neuro-Fuzzy Intelligent Systems. Springer, New York (2000)
5. Konar, A.: Computational Intelligence: Principles, Techniques and Applications. Springer (2005)
6. Korytkowski, M., Rutkowski, L., Scherer, R.: On speeding up the learning process of neuro-fuzzy ensembles generated by the adaboost algorithm. In: Kurzynski, M., Puchala, E., Wozniak, M., Zolnierek, A. (eds.) Computer Recognition Systems 2, pp. 318–326. Springer, Heidelberg (2007)
7. Korytkowski, M., Scherer, R., Rutkowski, L.: On combining backpropagation with boosting. In: 2006 International Joint Conference on Neural Networks, IEEE World Congress on Computational Intelligence, Vancouver, BC, Canada, pp. 1274–1277 (2006), doi:10.1109/IJCNN.2006.246838
8. Kuncheva, L.: Combining Pattern Classifiers. STUDFUZZ. John Wiley & Sons (2004)
9. Liu, Y., Yao, X.: Ensemble learning via negative correlation. Neural Networks 12, 1399–1404 (1999)
10. Liu, Y., Yao, X.: Simultaneous training of negatively correlated neural networks in an ensemble. IEEE Trans. Syst., Man, Cybern. B 29, 716–725 (1999)
11. Meir, R., Rätsch, G.: An Introduction to Boosting and Leveraging. In: Mendelson, S., Smola, A.J. (eds.) Advanced Lectures on Machine Learning. LNCS (LNAI), vol. 2600, pp. 118–183. Springer, Heidelberg (2003)
12. Schapire, R.E.: A brief introduction to boosting. In: Conference on Artificial Intelligence, pp. 1401–1406 (1999)
13. Setness, M., Babuska, R.: Bagging predictors. Machine Learning 26(2), 123–140 (1996)
14. Wang, L.X.: Adaptive Fuzzy Systems and Control. PTR Prentice-Hall, Englewood Cliffs (1994)

Chapter 4
Relational Modular Fuzzy Systems

This chapter presents the fuzzy relational model. In such model we define all possible connections between input and output linguistic terms [6, 12]. An advantage of this approach is great flexibility of the system. Input and output terms are fully interconnected. Moreover, the connections can be modeled by changing the elements of the relation matrix. The relation matrix can be regarded as a set of elements similar to rule weights in classic fuzzy systems [8, 11]. Relational fuzzy systems are used successfully to e.g. control [4] and classification tasks [1, 21, 23]. In this chapter, relational neuro-fuzzy systems [17, 20] will be used. Such neural network like structures allow to use more scenarios than in the case of ordinary relational structures. For example, we can set fuzzy linguistic values in advance and then fine tune the model mapping by changing relation elements using gradient learning. Gradient learning is an important advantage of relational neuro-fuzzy systems comparing to original fuzzy relational systems. Furthermore, this chapter presents the AdaBoost ensembles of relational neuro-fuzzy classifiers. A serious drawback of fuzzy system boosting ensembles is that such ensembles contain separate rule bases which cannot be directly merged. As systems are separate, we cannot treat fuzzy rules coming from different systems as rules from the same (single) system. The problem is addressed by a novel design of fuzzy systems constituting the ensemble, resulting in normalization of individual rule bases during learning. There were some attempts to combine fuzzy models, e.g. [13] or rough-fuzzy models [9] but none of them solved the problem of multiple rule bases in the ensemble.

4.1 Basic Relational Fuzzy Model

Relational fuzzy systems store associations between the input and the output linguistic values in the form of a discrete fuzzy relation. In a general case of a multi input multi output system (MIMO), the relation R is a multidimensional matrix containing degree of connection for every possible combination of input and output fuzzy sets. In a multi input single output (MISO), there are N inputs x_n, and a

R. Scherer: Multiple Fuzzy Classification Systems, STUDFUZZ 288, pp. 39–50.
springerlink.com

single output. To every input variable x_n we assign a set A_n of K_n linguistic values A_n^k, $k = 1,...,K_n$

$$A_n = \left\{A_n^1, A_n^2, ..., A_n^{K_n}\right\}. \tag{4.1}$$

To output variable y we assign a set B of M linguistic values B^m, $m = 1,...,M$, with membership functions $\mu_{B^m}(y)$, i.e.

$$B = \left\{B^1, B^2, ..., B^M\right\}. \tag{4.2}$$

In a case of large number of inputs in MIMO and MISO systems, the dimensionality of matrix R becomes quite high and it is very hard to determine elements of R. Therefore in the sequel we consider fuzzy systems with multidimensional input linguistic values. Then, we have only one set A of n-dimensional fuzzy linguistic values

$$A = \left\{A^1, A^2, ..., A^K\right\}. \tag{4.3}$$

Thus the relational matrix is only two dimensional in the MISO case. Sets A and B are related to each other with a certain degree by the $K \times M$ relation matrix

$$R = \begin{bmatrix} r_{11} & r_{12} & \cdots & r_{1M} \\ r_{21} & r_{22} & \cdots & r_{2M} \\ \vdots & \vdots & & \vdots \\ r_{K1} & r_{K2} & \cdots & r_{KM} \end{bmatrix}; \tag{4.4}$$

where element $r_{km} \in [0,1]$, $k = 1,...,K$, $m = 1,...,M$. Having given vector $\bar{A}$ of K membership values $\mu_{A^k}(\bar{\mathbf{x}})$ for a crisp observed input value $\bar{\mathbf{x}}$, vector $\bar{B}$ of M crisp memberships μ_m is obtained through a fuzzy relational composition

$$\bar{B} = \bar{A} \circ R, \tag{4.5}$$

implemented element-wise by a generalized form of t-conorm and t-norm composition, i.e. s-t composition

$$\mu_m = \underset{k=1}{\overset{K}{\mathrm{S}}} \left[\mathrm{T}\left(\mu_{A^k}(\bar{\mathbf{x}}), r_{km}\right)\right], \tag{4.6}$$

where T and S denote, respectively, t-norm and t-conorm. It should be noted that any triangular norms can be used in the s-t composition, however in the experiments the most common ones will be used – product and maximum. The crisp output of the relational system is computed by the weighted mean

$$\bar{y} = \frac{\sum\limits_{m=1}^{M} \left\{ \bar{y}^m \underset{k=1}{\overset{K}{\mathrm{S}}} \left[\mathrm{T}\left(\mu_{A^k}(\bar{\mathbf{x}}), r_{km}\right)\right] \right\}}{\sum\limits_{m=1}^{M} \underset{k=1}{\overset{K}{\mathrm{S}}} \left[\mathrm{T}\left(\mu_{A^k}(\bar{\mathbf{x}}), r_{km}\right)\right]}, \tag{4.7}$$

where $\bar{y}^m$ is a centre of gravity (centroid) of the fuzzy set B^m. The exemplary neuro-fuzzy structure of the relational system is depicted in Fig. 4.1. The first layer of the system consists of K multidimensional fuzzy membership functions. The second layer is responsible for the s-t composition of membership degrees from the previous layer and KM crisp numbers from the fuzzy relation. Finally, the third layer realizes center average defuzzification. Interpreting the system as a net structure allows learning or fine-tuning system parameters through the backpropagation algorithm. T-conorm in (4.7) can be replaced by ordered weighted averaging operators [25, 26] which further extends the versatility of the relational neuro-fuzzy system. The model described above can be a base to derive fuzzy rules which in the case of SISO (single input single output) relational model have the following form

$$\begin{aligned} &R^k : \text{ IF } \mathbf{x} \text{ is } A^k \text{ THEN} \\ &y \text{ is } B^1\left(r_{k1}\right), y \text{ is } B^m\left(r_{km}\right), \ldots \\ &\ldots, y \text{ is } B^M\left(r_{kM}\right), \end{aligned} \tag{4.8}$$

where r_{km} is a weight, reflecting the strength of connection between input and output fuzzy sets.

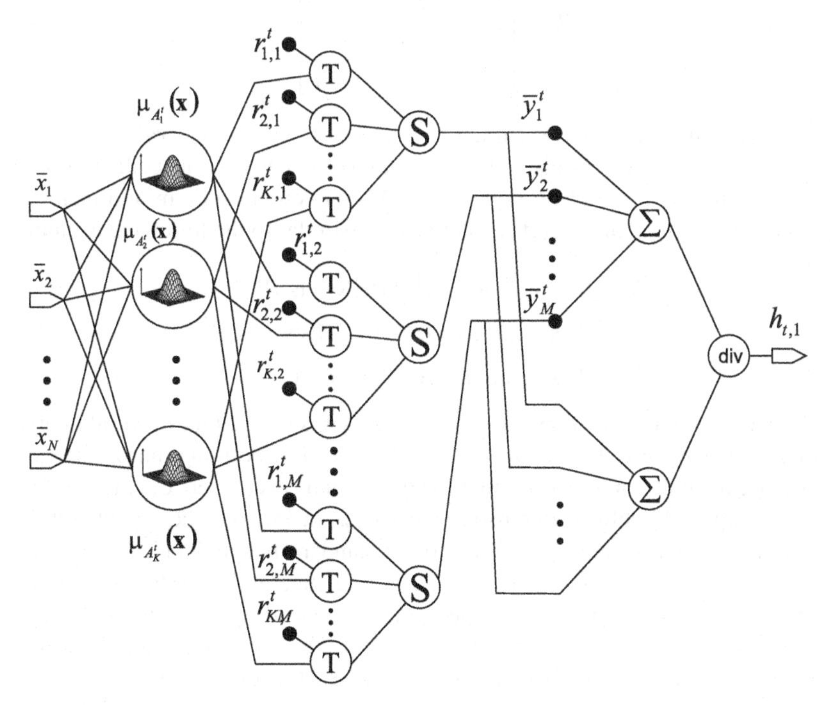

Fig. 4.1 Neuro-fuzzy relational system.

4.2 Generalized Relational Fuzzy Model

The power of fuzzy systems stems from their ability to process natural language expressions. We can model nearly any term using different shapes of fuzzy sets and various modifiers, i.e. fuzzy hedges. They transform original fuzzy linguistic values (primary terms e.g. "small", "big") into new, more specific fuzzy sets like "very fast" or "quite slow" flow. Thanks to them, we can make expert statements more precise. The concept of fuzzy hedges is well established in the literature. In general, they can be divided into powered and shifted hedges. Powered hedges model adverbs e.g. "very" by powering membership functions, whereas shifted hedges model adverbs by moving points of membership functions. In this section, a different view on linguistic hedges is presented. Instead of modifying antecedent or consequent linguistic values, additional fuzzy sets are introduced. In this approach, a fuzzy relational system with linguistic values defined on a unitary interval is used. These values are elements of a fuzzy relation matrix R connecting antecedent and consequent linguistic values. In this case, the relation matrix contains fuzzy sets C_{km} defined on a unitary interval

$$\mathbf{R} = \begin{bmatrix} C_{11} & C_{11} & \cdots & C_{1M} \\ C_{21} & C_{22} & \cdots & C_{2M} \\ \vdots & \vdots & C_{km} & \vdots \\ C_{K1} & C_{K2} & \cdots & C_{KM} \end{bmatrix} . \tag{4.9}$$

Thus, if we define several fuzzy linguistic values on unitary interval (e.g. see Fig. 4.2), an expert can express his or her uncertainty concerning antecedent terms by a linguistic description. In SISO systems, or MISO systems with multidimensional antecedent fuzzy sets, the expert can define rules similar to the following exemplary ones

$$\begin{aligned} R^1 &: \text{IF } \mathbf{x} \text{ is} \, exactly \, A^1 \text{ THEN } y \text{ is } B^1 \\ R^2 &: \text{IF } \mathbf{x} \text{ is} \, more \, or \, less \, A^1 \text{ THEN } y \text{ is } B^2 \\ R^3 &: \text{IF } \mathbf{x} \text{ is} \, roughly \, A^1 \text{ THEN } y \text{ is } B^3 \end{aligned} \tag{4.10}$$

The membership degree of an antecedent fuzzy set is divided into several intervals by fuzzy sets C_{km}, $k = 1,...,K$, $m = 1,...,M$. Instead of defining many antecedent sets we use a smaller number of input fuzzy sets and several sets C_{km}. Every fuzzy set A^k has up to M defined linguistic values C_{km}. In Fig. 4.2 there is also the set "not at all", which meaning is similar to the standard hedge "not". It is activated when its input fuzzy set A^k is not active. The inference in this system is similar to the sup-min composition, but min operation is replaced by a membership degree $\mu_{C_{km}}(\tau^k)$, where τ^k is the membership degree of the k-th multivariate input fuzzy set. The vector of crisp memberships, for $m = 1,...,M$, is obtained by

$$\mu_m = \underset{k=1}{\overset{K}{\mathrm{S}}} \left[\mu_{C_{km}} \left(\mu_{A^k} (\bar{\mathbf{x}}) \right) \right] . \tag{4.11}$$

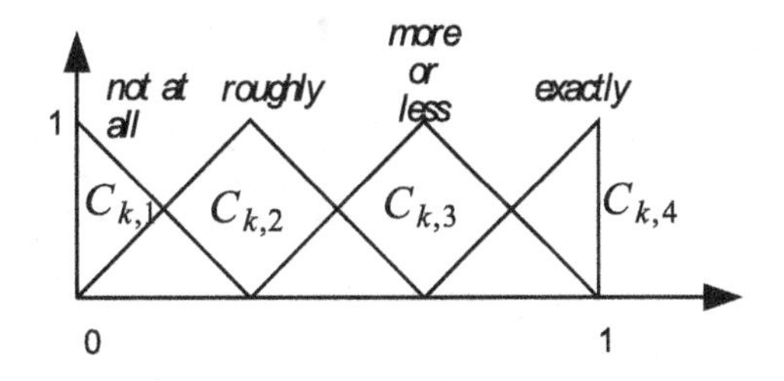

Fig. 4.2 Example of fuzzy linguistic values, expressing uncertainty in rule antecedents.

Eq. (4.11) reflects fuzzy hedge modifier operation. For example instead of quadratic function for concentration operation "very", we use a fuzzy set "exactly" (Fig. 4.2). Interpretation and operation of the sets in Fig. 4.2 is different from standard linguistic hedges. For example, standard fuzzy hedge "more or less" dilates an input fuzzy set, whereas our "roughly", "more or less" divide membership degree range into several intervals. The overall system output is computed through a weighted average

$$\bar{y} = \frac{\sum_{m=1}^{M} \left\{ \bar{y}^{m} \mathop{\mathrm{S}}_{k=1}^{K} \left[\mu_{C_{km}} \left(\mu_{A^{k}} \left(\bar{\mathbf{x}} \right) \right) \right] \right\}}{\sum_{m=1}^{M} \mathop{\mathrm{S}}_{k=1}^{K} \left[\mu_{C_{km}} \left(\mu_{A^{k}} \left(\bar{\mathbf{x}} \right) \right) \right]} . \tag{4.12}$$

The example is based on rules (4.10). The neuro-fuzzy structure of the new system is presented in Fig. 4.3). The first layer consists of K input multivariate fuzzy

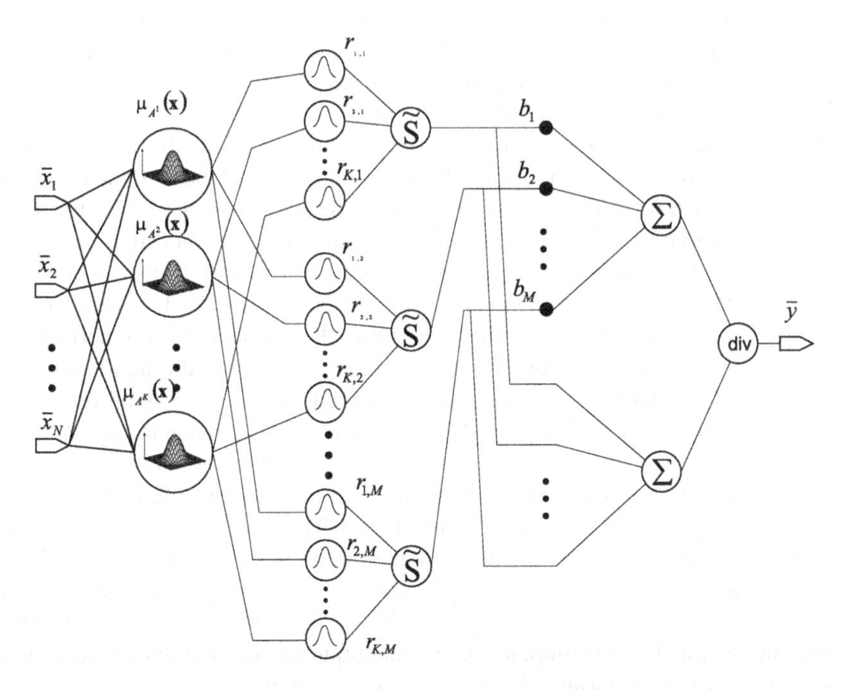

Fig. 4.3 The neuro-fuzzy structure of the relational system with fuzzy certainty degrees

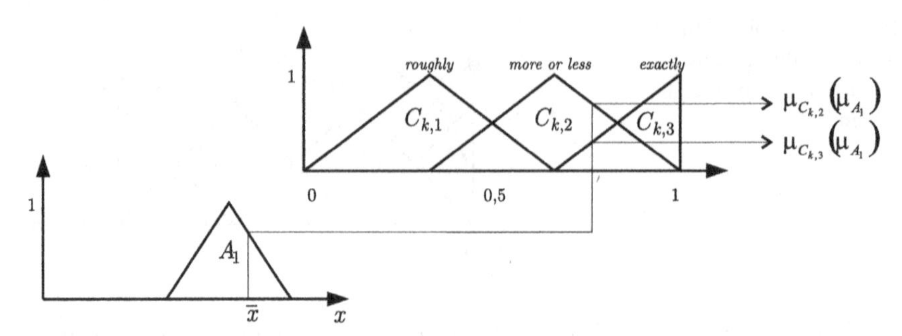

Fig. 4.4 Graphical inference of the new system on the basis of the exemplary rules (4.10)

membership functions. The second layer is responsible for the composition of certainty values and membership values from the previous layer. Finally, the third layer defuzzifies M output values through the center average method. All parameters of the neuro-fuzzy system can be tuned by the backpropagation algorithm.

4.3 Relational Boosting Ensemble

This section describes a method to build an ensemble of relational neuro-fuzzy systems with the possibility to merge all rule bases. To build the ensemble we use the AdaBoost algorithm. The algorithm assigns weights to learning samples according to their performance on earlier classifiers in the ensemble. Thus subsystems are trained with different datasets created from the base dataset. The most important problem in the case of creating ensembles from fuzzy systems as base hypothesis is that each rule base has different sum of rule activation levels. Thus we cannot treat them as one large fuzzy rule base. In other words we cannot disregard internal structure of the fuzzy rules arranged in fuzzy systems. To make use of the obtained rules it would be desirable to have a set of fuzzy rules which are not bound together by boosting output (3.9). In other words, subsystems from the boosting ensemble are bound together. For example, several relational neuro-fuzzy systems were connected together in an ensemble of classifiers [15] but rules in such system cannot be treated without thinking of the whole ensemble. This possible lack of normalization of rule bases comes from different sums of activation levels during training. To overcome this problem, this section proposes a method to normalize all relational rule bases during learning. The normalization is achieved by adding the second output to the system and keeping all rule bases at the same level. Using the proposed approach, we obtain a set of rules which can be treated as a single rule base that can be further processed (simplified, pruned etc.). The structure of relational systems will be modified to keep rule bases of all systems at same activity level (firing strength). The modification will consist in adding the second, normalising output. The AdaBoost algorithm is the most popular boosting method [5, 10, 14] and is described in Section 3.2. To compute the overall output of the ensemble of classifiers trained by AdaBoost algorithm the following formula is used

$$f(\mathbf{x}) = \sum_{t=1}^{T} c_t h_t(\mathbf{x}) \,, \tag{4.13}$$

where

$$c_t = \frac{\alpha_t}{\sum_{t=1}^{T} \alpha_t} \tag{4.14}$$

is classifier importance for a given training set, $h_t(\mathbf{x})$ is the response of the hypothesis t on the basis of feature vector $\mathbf{x} = [x_1,...,x_n]$. The coefficient c_t value is computed on the basis of the classifier error and can be interpreted as the measure of classification accuracy of the given classifier. As we see, the AdaBoost algorithm is a meta-learning algorithm and does not determine the way of learning for classifiers in the ensemble. In our approach, relational neuro-fuzzy systems are used as the hypothesis constituting the ensemble. The output of the single relational neuro-fuzzy system defined by (4.7), shown in Fig. 4.1, is

$$h_t = \frac{\sum_{m=1}^{M} \left\{ \bar{y}^m \mathop{\mathrm{S}}_{k=1}^{K} \left[\mathrm{T}\left(\mu_{A^k}(\bar{\mathbf{x}}), r_{km} \right) \right] \right\}}{\sum_{m=1}^{M} \mathop{\mathrm{S}}_{k=1}^{K} \left[\mathrm{T}\left(\mu_{A^k}(\bar{\mathbf{x}}), r_{km} \right) \right]} \,, \tag{4.15}$$

where index t, $t = 1,...,T$, of the classifier number is omitted for clarity. After putting (4.15) into (4.13) we obtain the final output of the classifier ensemble

$$f(\mathbf{x}) = \sum_{t=1}^{T} \frac{c_t \sum_{m=1}^{M} \left\{ \bar{y}^m \mathop{\mathrm{S}}_{k=1}^{K} \left[\mathrm{T}\left(\mu_{A^k}(\bar{\mathbf{x}}), r_{km} \right) \right] \right\}}{\sum_{m=1}^{M} \mathop{\mathrm{S}}_{k=1}^{K} \left[\mathrm{T}\left(\mu_{A^k}(\bar{\mathbf{x}}), r_{km} \right) \right]} \tag{4.16}$$

Formula (4.16) is the weighted sum of all hypothesis outcomes. It is not possible to merge all rule bases coming from members of the ensemble, because (4.16) cannot be rewritten to use a common denominator. Let us observe that in the denominator of (4.16) there is the sum of activity levels of rules in a single neuro-fuzzy system. Thus if we want to treat formula (4.16) as one neuro-fuzzy system, it should be transformed so that it has the sum of activity level of all rules in the ensemble. The solution to the problem is the following assumption for every module constituting the ensemble

$$\sum_{m=1}^{M} \mathop{\mathrm{S}}_{k=1}^{K} \left[\mathrm{T}\left(\mu_{A^k}(\bar{\mathbf{x}}), r_{km} \right) \right] = 1, \; \forall t = 1,...,T \tag{4.17}$$

Then we can get rid of the denominator in (4.16). The assumption (4.17) guarantees that in the case of creating an ensemble of several such modified relational neuro-fuzzy systems, activity level of one system does not dominate other subsystems. Unfortunately there does not exist a method for designing neuro-fuzzy systems so that the sum of the activity levels for a single system equals one. One possibility is such selection of fuzzy system parameters during learning that assumption (4.17) is satisfied. Considered neuro-fuzzy systems should be trained to satisfy (4.17) and

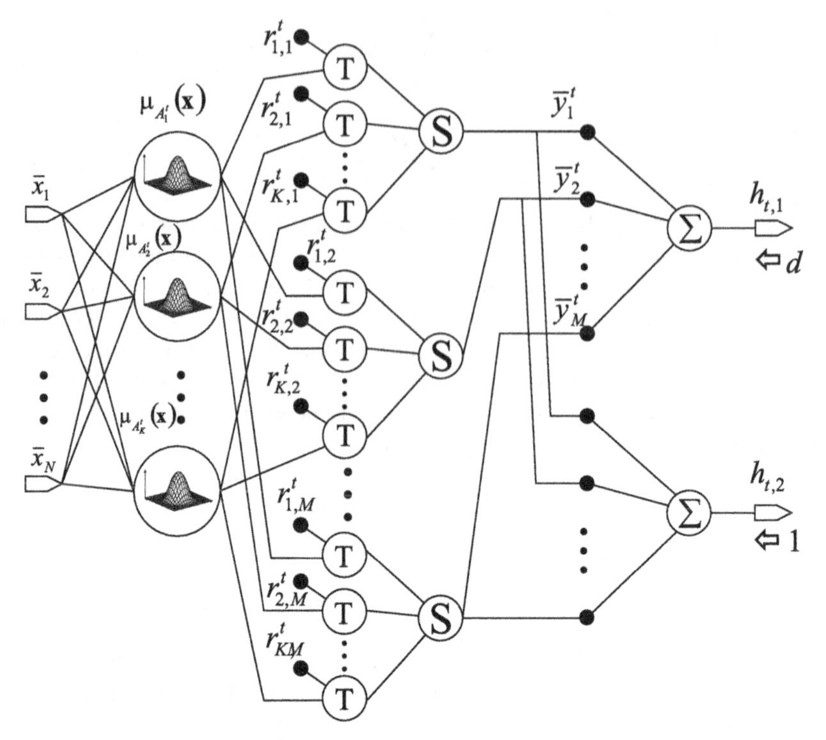

Fig. 4.5 Single modified relational neuro-fuzzy system. The system is a part of the boosting ensemble. The second output is learned with the constant desired response which equals 1.

to make the output being equal the desired value. To satisfy these requirements, we transform the neuro-fuzzy structure in Fig. 4.1 to obtain the form depicted in Fig. 4.5. We removed the last layer performing the division, thus the system has two outputs. The error on the first output will be computed taking into account desired output from learning data. Desired signal on the second output is constant and equals 1. As we assume the fulfillment of the condition (4.17), after learning we can remove from the structure elements responsible for the denominator in (4.15), as the denominator equals 1. The single subsystem takes the form shown in Fig. 4.5. Rules from an ensemble of such trained neuro-fuzzy relational classifiers can be treated as one large fuzzy rules base.

4.4 Experimental Results

In this section we describe learning of the neuro-fuzzy relational systems and an experimental example. As aforementioned, the AdaBoost algorithm is a meta learning algorithm and has to be supported by some regular training. The systems were

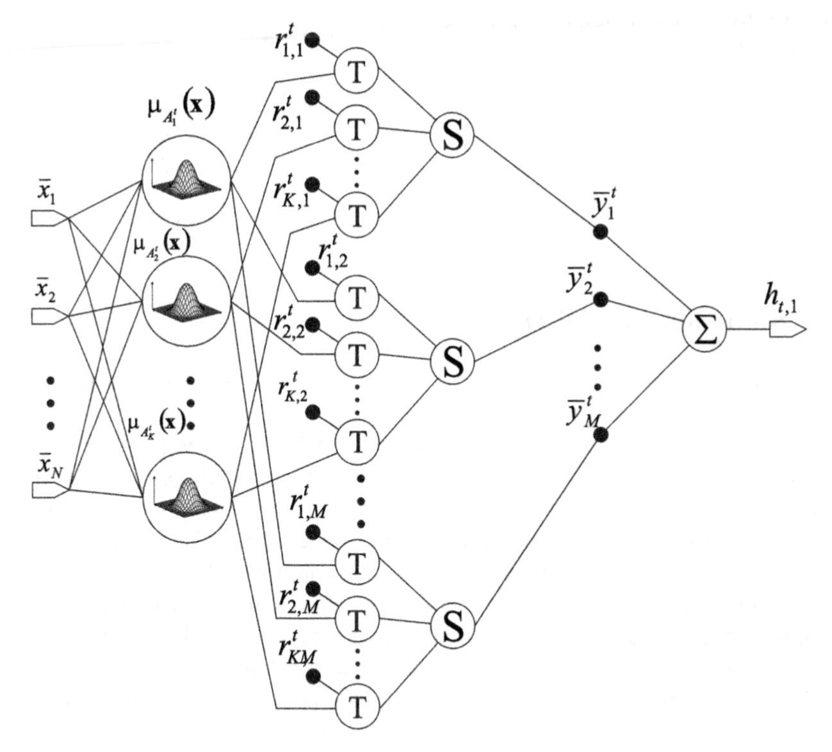

Fig. 4.6 Single modified relational neuro-fuzzy system after learning. The system is a part of the boosting ensemble. It was possible to remove the second output as it was needed only during the learning stage.

initialized by the modified fuzzy c-means algorithm [3, 2] (Section 3.5) and trained by the backpropagation algorithm [24] (Section 3.4). AdaBoost influences the backpropagation algorithm by changing learning rate accordingly to boosting weights. This is achieved by multiplication of the error value in the backpropagation algorithm by the weight of the current learning vector (Section 3.4). It should be emphasized that the goal of experiments was to illustrate the proposed approach and to show the ability of the systems to learn. We do not compare the obtained results with the results from the literature, as the aim of the experiments was to show the possibility to merge all systems from the ensemble into one rule base. The experiment was performed on the Glass Identification dataset [7] (see Section 1.2.1). We took out of 214 instances 43 instances for a testing set. We obtained 93.40% classification accuracy using relational NFS with 6 antecedent fuzzy sets and 3 consequent fuzzy sets. We used three relational neuro-fuzzy classifiers described in Table 4.1. Experiment showed good performance and convergence of the proposed approach, comparable to those from the literature.

Table 4.1 Experimental results of the Glass Identification problem on the AdaBoost ensemble of the relational neuro-fuzzy systems

	Classifier 1	Classifier 2	Classifier 3
No. of input/output sets	2/2	2/2	2/2
No. of epochs	20	20	25
Coefficients c_t	0.34	0.33	0.33
Classification %	97.5	95.31	92.4

4.5 Summary and Discussion

In this chapter we presented neuro–fuzzy relational models. Relational neuro-fuzzy systems are very versatile and can be used in various tasks of classification, modeling and prediction. They store associations between input and output fuzzy sets in the form discrete relation. Section 4.1 describes basic relational system. Section 4.2 presents a new fuzzy model with fuzzy sets in matrix 4.9 linking input and output fuzzy sets. These linking fuzzy sets are elements of the relation. Thus, we can model descriptions like IF x is *roughly* fast THEN .... This approach does not involve standard fuzzy hedges, so input and output linguistic values remain intact. This improves the transparency of the model. Numerical simulations showed reliable behaviour of the new system. Its performance is comparable with the popular singleton model, but the new system allows expressing expert's uncertainty regarding input linguistic variables.

Section 4.3 shows how to create an AdaBoost ensemble of relational fuzzy classifiers. The hypothesis within the ensemble are trained by the fuzzy c-means algorithm and the backpropagation algorithm. Thanks to inherent interpretability in fuzzy systems, it is easy to initialize the system using some expert knowledge or exploratory data analysis methods. Fuzzy rules coming from different fuzzy systems cannot be merged immediately. We proposed a method to overcome this disadvantage. The fuzzy classifiers are modified in a specific way to keep the activation level of each rule base at the same level. Thus none of classifiers dominates over the ensemble. The modification is very simple and does not introduce additional computational burden.

In the numerical simulations, linguistic values are set in advance by the fuzzy c-means algorithm. Then all system parameters (relation matrix elements and membership function parameters) are determined by the backpropagation algorithm. Simulations on some popular benchmarks show great accuracy of the proposed method. The obtained results are comparable to the best results in the literature. Our goal was to show the ability to learn and to obtain a set of fuzzy rules ready to be further processed, e.g. to prune and simplify the rule base. However, the most important advantage of the proposed method is the possibility of merging several separate fuzzy rule bases into one set of fuzzy rules.

The fundamental results of this chapter were partly presented in [18, 19, 20, 21, 16, 15, 22]. Further research could include rule extraction and simplification. It is also possible to apply proposed systems in various classification tasks and some of the systems can be used for nonlinear modeling and prediction.

References

1. Babuska, R.: Fuzzy Modeling For Control. Kluwer Academic Press, Boston (1998)
2. Bezdek, J., Keller, J., Krisnapuram, R., Pal, N.: Fuzzy Models and Algorithms for Pattern Recognition and Image Processing. Kluwer Academic Press (1999)
3. Bezdek, J.C., Pal, S.: Fuzzy Models for Pattern Recognition. IEEE Press, New York (1992)
4. Branco, P., Dente, J.: A fuzzy relational identification algorithm and its application to predict the behaviour of a motor drive system. Fuzzy Sets and Systems 109, 343–354 (2000)
5. Breiman, L.: Bias, variance, and arcing classifiers. Tech. Rep. Technical Report 460, Statistics Department, University of California (1997)
6. Ciaramella, A., Tagliaferri, R., Pedrycz, W., Di Nola, A.: Fuzzy Relational Neural Network for Data Analysis. In: Di Gesú, V., Masulli, F., Petrosino, A. (eds.) WILF 2003. LNCS (LNAI), vol. 2955, pp. 103–109. Springer, Heidelberg (2006)
7. Frank, A., Asuncion, A.: UCI machine learning repository (2010), `http://archive.ics.uci.edu/ml`
8. Ishibuchi, H., Nakashima, T.: Effect of rule weights in fuzzy rule-based classification systems. IEEE Trans. on Fuzzy Systems 9, 506–515 (2000)
9. Korytkowski, M., Nowicki, R., Scherer, R.: Neuro-fuzzy Rough Classifier Ensemble. In: Alippi, C., Polycarpou, M., Panayiotou, C., Ellinas, G. (eds.) ICANN 2009, Part I. LNCS, vol. 5768, pp. 817–823. Springer, Heidelberg (2009)
10. Meir, R., Rätsch, G.: An Introduction to Boosting and Leveraging. In: Mendelson, S., Smola, A.J. (eds.) Advanced Lectures on Machine Learning. LNCS (LNAI), vol. 2600, pp. 118–183. Springer, Heidelberg (2003)
11. Nauck, D., Kruse, R.: How the learning of rule weights affects the interpretability of fuzzy systems. In: Proceedings of 1998 IEEE World Congress on Computational Intelligence, FUZZ-IEEE, pp. 1235–1240 (1998)
12. Pedrycz, W.: Fuzzy Control and Fuzzy Systems. Research Studies Press, London (1989)
13. Pedrycz, W., Kwak, K.: Boosting of granular models. Fuzzy Sets and Systems 157, 2934–2953 (2006)
14. Schapire, R.E.: A brief introduction to boosting. In: Conference on Artificial Intelligence, pp. 1401–1406 (1999)
15. Scherer, R.: Boosting Ensemble of Relational Neuro-fuzzy Systems. In: Rutkowski, L., Tadeusiewicz, R., Zadeh, L.A., Żurada, J.M. (eds.) ICAISC 2006. LNCS (LNAI), vol. 4029, pp. 306–313. Springer, Heidelberg (2006)
16. Scherer, R.: Regression Modeling with Fuzzy Relations. In: Rutkowski, L., Tadeusiewicz, R., Zadeh, L.A., Zurada, J.M. (eds.) ICAISC 2008. LNCS (LNAI), vol. 5097, pp. 317–323. Springer, Heidelberg (2008), `http://www.springerlink.com/content/g3t810v513t4284w/`
17. Scherer, R.: Neuro-fuzzy relational systems for nonlinear approximation and prediction. Nonlinear Analysis 71, e1420–e1425 (2009), `http://linkinghub.elsevier.com/retrieve/pii/S0362546X09001898`, doi:10.1016/j.na.2009.01.18

18. Scherer, R.: Designing boosting ensemble of relational fuzzy systems. International Journal of Neural Systems 20(5), 381–388 (2010), http://www.worldscinet.com/ijns/20/2005/S0129065710002528.html
19. Scherer, R.: Neuro-fuzzy Systems with Relation Matrix. In: Rutkowski, L., Scherer, R., Tadeusiewicz, R., Zadeh, L.A., Zurada, J.M. (eds.) ICAISC 2010. LNCS (LNAI), vol. 6113, pp. 210–216. Springer, Heidelberg (2010), http://springerlink.com/content/px5gux3451512xg4/
20. Scherer, R., Rutkowski, L.: Neuro-fuzzy relational systems. In: 2002 International Conference on Fuzzy Systems and Knowledge Discovery, Singapore, November 18-22, pp. 44–48 (2002)
21. Scherer, R., Rutkowski, L.: Neuro-Fuzzy Relational Classifiers. In: Rutkowski, L., Siekmann, J.H., Tadeusiewicz, R., Zadeh, L.A. (eds.) ICAISC 2004. LNCS (LNAI), vol. 3070, pp. 376–380. Springer, Heidelberg (2004)
22. Scherer, R., Rutkowski, L.: Connectionist fuzzy relational systems. In: Hagamuge, S., Wang, L. (eds.) Computational Intelligence for Modelling and Control. SCI, pp. 35–47. Springer (2005)
23. Setness, M., Babuska, R.: Fuzzy relational classifier trained by fuzzy clustering. IEEE Transactions on Systems, Man and Cybernetics - Part B: Cybernetics 29(5), 619–625 (1999)
24. Wang, L.X.: Adaptive Fuzzy Systems and Control. PTR Prentice-Hall, Englewood Cliffs (1994)
25. Yager, R., Filev, D.: Essentials of Fuzzy Modeling and Control. John Wiley & Sons Inc., New York (1994)
26. Yager, R., Filev, D.: On a flexible structure for fuzzy systems models. In: Yager, R., Zadeh, L. (eds.) Fuzzy Sets, Neural Networks, and Soft Computing, pp. 1–28. Van Nostrand Reinhold, New York (1994)

Chapter 5
Ensembles of the Mamdani Fuzzy Systems

This chapter describes a family of fuzzy systems that use neural network like approach for learning and visualizing the system. Models in this chapter have their antecedents and consequents of rules connected by a t-norm. Such systems are called the Mamdani type neuro-fuzzy systems and they are the most common neuro-fuzzy systems. As it is emphasized in the previous chapter, the most important problem in case of creating ensembles from fuzzy systems as base hypothesis is that each rule base has different overall activation level. Thus we cannot treat them as one large fuzzy rule base. This possible "inequality" comes from different activation level during training. To overcome this problem, we apply the method proposed in the previous chapter to normalize all rule bases during learning. The normalization is achieved by adding the second output to the Mamdani system and keeping all rule bases at the same level.

5.1 Mamdani Neuro-fuzzy Systems

The structure of such systems depends to a large extent on defuzzification type and in this chapter we consider systems with the following defuzzification

$$\overline{y} = \frac{\sum_{r=1}^{N} \overline{y}^r \cdot \mu_{\overline{B}^r}(\overline{y}^r)}{\sum_{r=1}^{N} \mu_{\overline{B}^r}(\overline{y}^r)}. \tag{5.1}$$

In such systems, at the inference block output we obtain N fuzzy sets. The membership functions of fuzzy sets $\overline{B}^r$, $r = 1,2,\ldots,N$, are defined using sup-t composition

$$\mu_{\overline{B}^r}(y) = \sup_{\mathbf{x}\in\mathbf{X}} \left\{ \mu_{A^r}(\mathbf{x}) \overset{T}{*} \mu_{A^r \to B^r}(\mathbf{x}, y) \right\}. \tag{5.2}$$

In case of typical, singleton type fuzzification, formula (5.2) takes the form

$$\mu_{\overline{B}^r}(y) = \mu_{A^r \to B^r}(\overline{\mathbf{x}}, y) = T\left(\mu_{A^r}(\overline{\mathbf{x}}), \mu_{B^r}(y)\right). \tag{5.3}$$

R. Scherer: Multiple Fuzzy Classification Systems, STUDFUZZ 288, pp. 51–59.
springerlink.com

Since

$$\mu_{A^r}(\overline{\mathbf{x}}) = \mathop{T}_{i=1}^{n}\left(\mu_{A_i^r}(\overline{x}_i)\right), \tag{5.4}$$

we have

$$\mu_{\overline{B}^r}(y) = T\left[\mathop{T}_{i=1}^{n}\left(\mu_{A_i^r}(\overline{x}_i)\right), \mu_{B^r}(y)\right], \tag{5.5}$$

where T is any t-norm. Since

$$T(a,1) = a \tag{5.6}$$

and

$$\mu_{B^r}(\overline{y}^r) = 1\,, \tag{5.7}$$

we obtain the following dependency:

$$\mu_{\overline{B}^r}(\overline{y}^r) = \mathop{T}_{i=1}^{n}\left(\mu_{A_i^r}(\overline{x}_i)\right). \tag{5.8}$$

By substituting formula (5.8) to formula (5.1), we get

$$\overline{y} = \frac{\sum_{r=1}^{N} \overline{y}^r \cdot T_{i=1}^{n}\left(\mu_{A_i^r}(\overline{x}_i)\right)}{\sum_{r=1}^{N} T_{i=1}^{n}\left(\mu_{A_i^r}(\overline{x}_i)\right)}. \tag{5.9}$$

In such systems, separate inference is made within each rule and $\mu_{\overline{B}^r}(\overline{y}^r)$, $r = 1,2,\ldots,N$, is computed. Let us assume that input and output linguistic variables are described by means of Gaussian membership functions, that is

$$\mu_{A_i^r}(x_i) = \exp\left[-\left(\frac{x_i - \overline{x}_i^r}{\sigma_i^r}\right)^2\right], \tag{5.10}$$

$$\mu_{B^r}(y) = \exp\left[-\left(\frac{y - \overline{y}^r}{\sigma^r}\right)^2\right]. \tag{5.11}$$

By substituting the above dependencies to formula (5.8) and applying the t-norm, we will get the following description:

$$\overline{y} = \frac{\sum_{r=1}^{N} \overline{y}^r \left(\prod_{i=1}^{n} \exp\left[-\left(\frac{\overline{x}_i - \overline{x}_i^r}{\sigma_i^r}\right)^2\right]\right)}{\sum_{r=1}^{N} \left(\prod_{i=1}^{n} \exp\left[-\left(\frac{\overline{x}_i - \overline{x}_i^r}{\sigma_i^r}\right)^2\right]\right)}. \tag{5.12}$$

Let us notice that in dependency (5.12), there is no parameter σ^r of the output fuzzy set B^r, $r = 1,2,\ldots,N$. The network–like structure of the system is shown in Fig. 5.1.

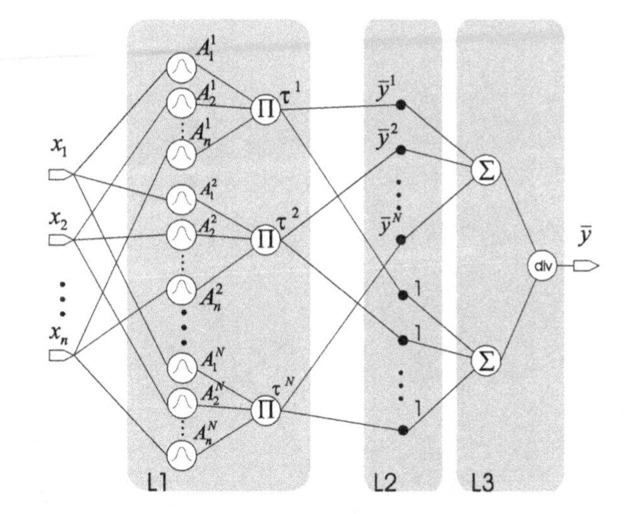

Fig. 5.1 Network structure of the system described by formula (5.12)

5.2 Interpretable Boosting Ensembles of the Mamdani Systems

This section describes the method to build an ensemble of the Mamdani neuro-fuzzy systems with the possibility to merge all rule bases. The Mamdani systems are used as the hypothesis constituting the ensemble trained by the AdaBoost algorithm [1, 7, 8], described in Section 3.1.

Analogously to the approach proposed in Section 4.3 , the structure of the Mamdani neuro-fuzzy system will be modified to keep rule bases of all systems at same activity level (firing strength). The modification will consist in adding the second, normalising output.

The output of the single Mamdani system defined by (5.12) is

$$h_t = \frac{\sum\limits_{r=1}^{N_t} \bar{y}_t^r \cdot \tau_t^r}{\sum\limits_{r=1}^{N_t} \tau_t^r} \tag{5.13}$$

where $\tau_t^r = \mathop{T}\limits_{i=1}^{n} \left(\mu_{A_i^r}(\bar{x}_i) \right)$ is rule activation level, index $t = 1, ..., T$ of the classifier number is omitted for clarity. After putting (5.13) into (3.9) we obtain the final output of the classifier ensemble

$$f(\mathbf{x}) = \sum_{t=1}^{T} c_t \frac{\sum\limits_{r=1}^{N_t} \bar{y}_t^r \cdot \tau_t^r}{\sum\limits_{r=1}^{N_t} \tau_t^r} \tag{5.14}$$

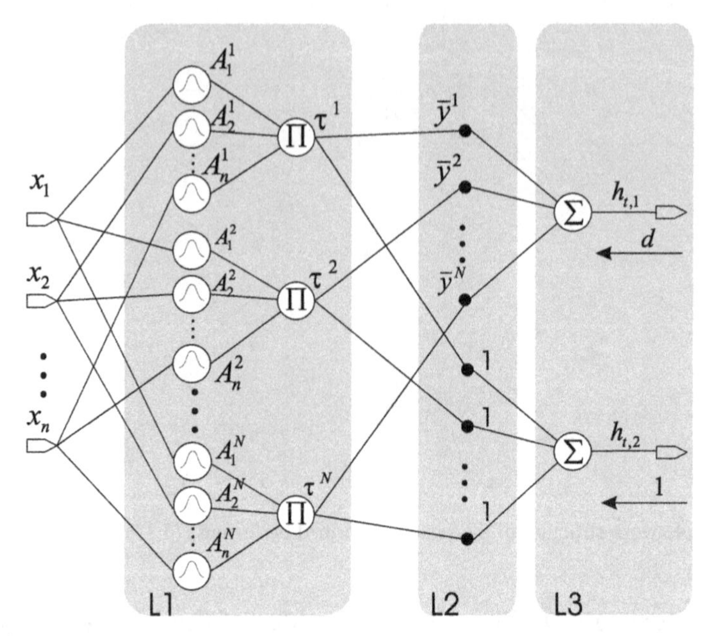

Fig. 5.2 The single modified Mamdani neuro-fuzzy system. The system is a part of the boosting ensemble. The second output is learned with the constant desired response which equals 1.

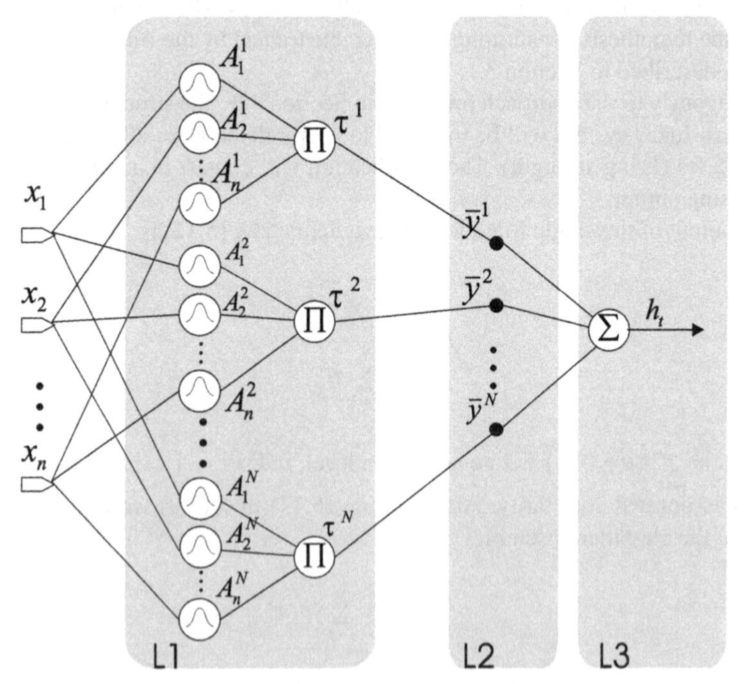

Fig. 5.3 Single modified relational neuro-fuzzy system after learning. The system is a part of the boosting ensemble. It was possible to remove the second output as it was needed only during the learning stage.

As formula (5.14) is the weighted sum of all hypothesis, we cannot merge rule bases of all ensemble members. Analogously to Section 4.3, we will assume for every subsystem that

$$\sum_{r=1}^{N_t} \tau_t^r = 1 \ \forall t = 1, ..., T \tag{5.15}$$

Using (5.15) during learning, we can remove the denominator in (5.14). To satisfy (5.15), we transform the neuro-fuzzy structure in Fig. 5.1 to obtain the form presented in Fig. 5.2. We removed the last layer performing the division, thus the system has two outputs. The error on the first output will be computed taking into account desired output from learning data. Desired signal on the second output is constant and equals 1. As we assume the fulfillment of the condition (5.15), after learning we can remove from the structure elements responsible for the denominator in (4.15), and the system takes the form shown in Fig. 5.3. After learning consecutive structures according to the proposed idea, we can build the modular system presented in Fig. 5.4, in which the rules can be arranged in arbitrary order. We do not have to remember from which submodule the rule comes from. The obtained ensemble of neuro-fuzzy systems can be interpreted as a regular single neuro-fuzzy system, where coefficients c_t (see Eq. 3.10) can be interpreted as weights of fuzzy rules. Important ability of the system is its possible fine tuning. Yet in order to make the structure acting like a neuro-fuzzy system during learning, we have to add the

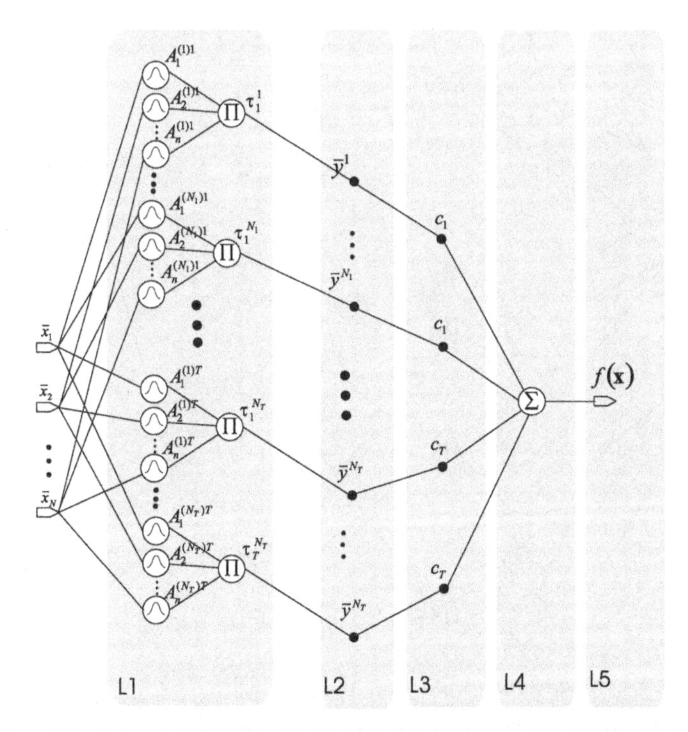

Fig. 5.4 Ensemble of the modified Mamdani neuro-fuzzy systems combined after learning.

removed elements responsible for denominator in (5.13). Alternatively we can use again the modification proposed in Section 3.4 for learning. The system will be similar to that shown in Figure 5.2.

5.3 Negative Correlation Learning Ensembles of the Mamdani Systems

This section presents an application of the negative correlation learning [5, 6], which is a meta learning algorithm for creating negatively correlated ensembles (see Section 3.3), to an ensemble of the Mamdani type neuro-fuzzy systems. All system

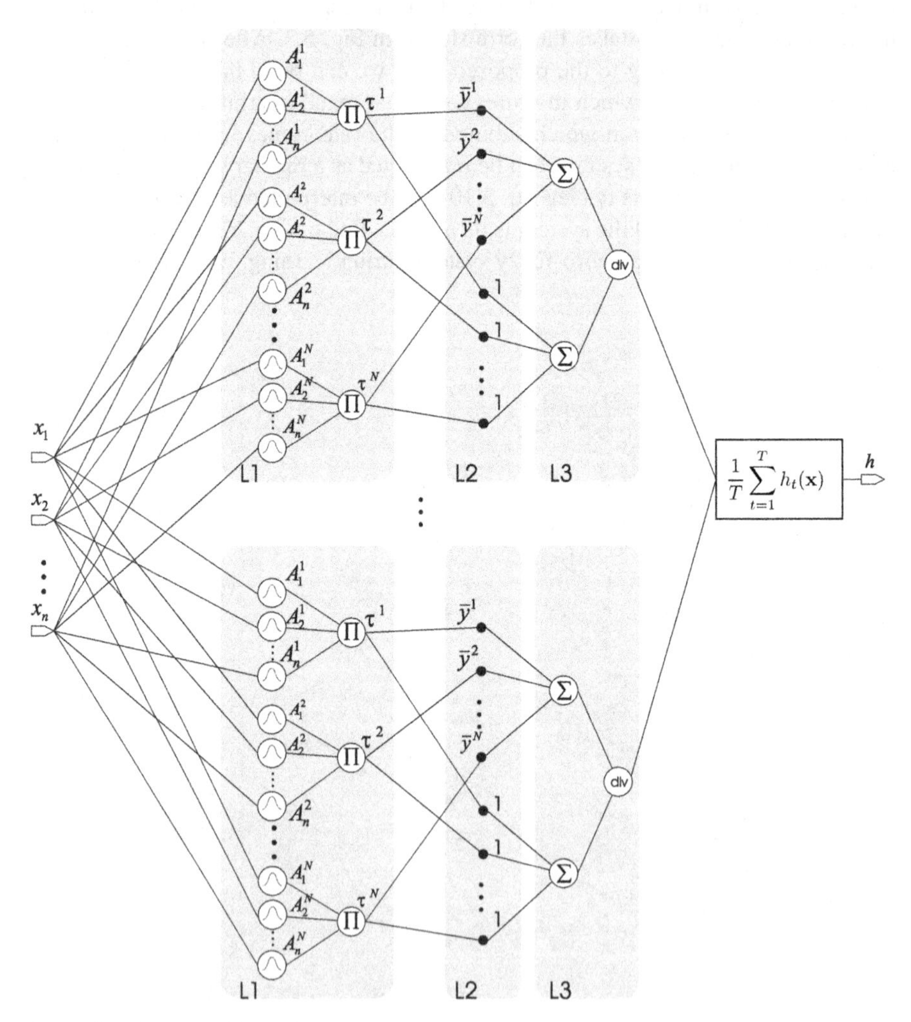

Fig. 5.5 NCL ensemble of the Mamdani neuro-fuzzy systems.

parameters were determined by the backpropagation algorithm. The NCL penalty factor influenced the backpropagation learning by inhibiting fuzzy system parameters modification if the system was correlated with the ensemble. The ensemble is shown in Figure 5.5. Experiments on some popular benchmarks show great accuracy of the proposed method.

5.4 Experimental Results

In this section we describe learning of the neuro-fuzzy systems and simulation examples. Knowledge of neuro-fuzzy systems is expressed in the form of fuzzy rules. During learning, the right number of rules, fuzzy set parameters and weights are to be determined. Alternatively they can be set by an expert. In our simulations we compute all the parameters by machine learning from numerical data, that is by the AdaBoost algorithm (Section 3.2) and by the backpropagation algorithm (Section 3.4). At the beginning, input fuzzy sets are determined by the modified fuzzy c-means clustering algorithm (Section 3.5). Then, all parameters are tuned by the backpropagation algorithm. Having given learning data set of pair $(\bar{\mathrm{x}}, d)$, where d is the desired response of the system, we can use the learning error to compute all parameters. Signal d is used as the desired response of the first output and the second output is learned with the desired response which equals 1. All experiments in this chapter are performed on ensembles of the modified Mamdani neuro-fuzzy systems with the additional output (see Fig. 5.2). The first experiments was carried out on the Wisconsin breast cancer database [2] (Section 1.2.4). From the data set 205 instances were taken into testing data and 16 instances with missing features were removed. The classification accuracy is 97,45%. Detailed errors and subsystem parameters are described in Table 5.1. The second experiment was performed on the Glass Identification dataset [2] (see Section 1.2.1). The goal is to classify 214 instances of glass into window and non-window glass basing on 9 numeric features. We took out 43 instances for a testing set. We obtained 97.34% classification accuracy. We used 5 neuro-fuzzy classifiers described in Table 5.2. Simulations shows superior performance and faster convergence of the proposed approach comparing to those from the literature. We simulated also the NCL ensemble of the Mamdani systems (Section 3.3). The systems were initialized randomly by the fuzzy c-means clustering and then trained the backpropagation algorithm. From the WBCD data set 205 instances were taken into testing data and 16 instances with missing features were removed. The classification accuracy was 95.3%. In case of the Glass Identification problem we took out 43 instances for a testing set. We obtained 94.23%

Table 5.1 Numerical results of Wisconsin breast cancer database

	Classifier 1	Classifier 2	Classifier 3
No. of epochs	15	50	30
Coefficients c_t	0.29	0.35	0.36
Classification %	97.42	97.45	97.41

Table 5.2 Numerical results of Glass Identification

	Classifier 1	Classifier 2	Classifier 3	Classifier 4	Classifier 5
No. of rules	2	2	2	2	2
No. of epochs	30	30	30	50	50
Coefficients c_t	0.13	0.01	0.18	0.33	0.33
Classification %	89.34	77.23	89.8	97.34	97.53

classification accuracy. We used 3 neuro-fuzzy classifiers. In the Ionosphere problem [2] (see Section 1.2.2) the data were divided randomly into 246 learning and 105 testing instances. We used 4 neuro-fuzzy classifiers and we obtained 94.03% classification accuracy.

5.5 Summary and Discussion

Neuro-fuzzy systems have numerous advantages over neural networks such as the possibility to compute their initial state and interpretable knowledge. To improve their performance we connected several fuzzy systems into an ensemble of classifiers. The ensembles in this chapter were created using the AdaBoost algorithm and NCL. The fuzzy subsystems are learned by gradient learning and initialized by modified fuzzy c-means clustering algorithm. To make possible merging fuzzy rule bases coming from the AdaBoost subsystems, we modify them by adding special second output. Thanks to the desired signal on this output being equal 1, it is possible to remove it afterward and merge the systems as none of them dominate over the rest. Experiments on well known benchmarks show very good ability to classify data. Results are comparable to the best ones from the literature. The fundamental results of this chapter were partly presented in [3, 4].

References

1. Breiman, L.: Bias, variance, and arcing classifiers. Tech. Rep. Technical Report 460, Statistics Department, University of California (1997)
2. Frank, A., Asuncion, A.: UCI machine learning repository (2010), http://archive.ics.uci.edu/ml
3. Korytkowski, M., Rutkowski, L., Scherer, R.: From Ensemble of Fuzzy Classifiers to Single Fuzzy Rule Base Classifier. In: Rutkowski, L., Tadeusiewicz, R., Zadeh, L.A., Zurada, J.M. (eds.) ICAISC 2008. LNCS (LNAI), vol. 5097, pp. 265–272. Springer, Heidelberg (2008), http://www.springerlink.com/content/p2813172x2w414t2/
4. Korytkowski, M., Scherer, R.: Negative Correlation Learning of Neuro-fuzzy System Ensembles. In: Rutkowski, L., Scherer, R., Tadeusiewicz, R., Zadeh, L.A., Zurada, J.M. (eds.) ICAISC 2010. LNCS (LNAI), vol. 6113, pp. 114–119. Springer, Heidelberg (2010), http://www.springerlink.com/content/n4732566203326q1/
5. Liu, Y., Yao, X.: Ensemble learning via negative correlation. Neural Networks 12, 1399–1404 (1999)

6. Liu, Y., Yao, X.: Simultaneous training of negatively correlated neural networks in an ensemble. IEEE Trans. Syst., Man, Cybern. B 29, 716–725 (1999)
7. Meir, R., Rätsch, G.: An Introduction to Boosting and Leveraging. In: Mendelson, S., Smola, A.J. (eds.) Advanced Lectures on Machine Learning. LNCS (LNAI), vol. 2600, pp. 118–183. Springer, Heidelberg (2003)
8. Schapire, R.E.: A brief introduction to boosting. In: Conference on Artificial Intelligence, pp. 1401–1406 (1999)
9. Wang, L.X.: Adaptive Fuzzy Systems and Control. PTR Prentice-Hall, Englewood Cliffs (1994)

Chapter 6
Logical Type Fuzzy Systems

The previous chapter described the Mamdani neuro-fuzzy systems which are the most common neuro-fuzzy systems. This chapter presents systems with a fuzzy implication connecting the antecedents and the consequents of fuzzy rules. Such systems are proved to perform better in classification tasks [5].

6.1 Neuro-fuzzy Systems of Logical Type

In logical type systems, the defuzzification is made by means of the following formula (see Section 2.5.4)

$$\overline{y} = \frac{\sum_{r=1}^{N} \overline{y}^r \cdot \mu_{B'}\left(\overline{y}^r\right)}{\sum_{r=1}^{N} \cdot \mu_{B'}\left(\overline{y}^r\right)}. \tag{6.1}$$

In these systems, the fuzzy set B' is created as a result of intersection of fuzzy sets $\overline{B}^k$, i.e.

$$B' = \bigcap_{k=1}^{N} \overline{B}^k. \tag{6.2}$$

The membership function of fuzzy set B' is determined using a t-norm, which shall be notated as follows:

$$\mu_{B'}(y) = \mathop{T}_{k=1}^{N} \left\{\mu_{\overline{B}^k}(y)\right\}. \tag{6.3}$$

Using formulas (6.1), (6.2) and (6.3), we have

$$\begin{aligned} \mu_{B'}\left(\overline{y}^r\right) &= \mathop{T}_{k=1}^{N} \left\{\mu_{\overline{B}^k}\left(\overline{y}^r\right)\right\} = \mathop{T}_{k=1}^{N} \left\{I\left(\mu_{A^k}\left(\overline{\mathbf{x}}\right), \mu_{B^k}\left(\overline{y}^r\right)\right)\right\} \\ &= \mathop{T}_{k=1}^{N} \left\{I\left(\mathop{T}_{i=1}^{N} \mu_{A_i^k}\left(\overline{x}_i\right), \mu_{B^k}\left(\overline{y}^r\right)\right)\right\}, \end{aligned} \tag{6.4}$$

where I is a fuzzy implication defined in section 2.5.4. By substituting formula (6.4) to dependency (6.1), we obtain

R. Scherer: Multiple Fuzzy Classification Systems, STUDFUZZ 288, pp. 61–71.
springerlink.com

$$\bar{y} = \frac{\sum_{r=1}^{N} \bar{y}^r \cdot T_{k=1}^{N} \left\{ I\left(T_{i=1}^{n} \left\{ \mu_{A_i^k}(\bar{x}_i) \right\}, \mu_{B^k}(\bar{y}^r) \right) \right\}}{\sum_{r=1}^{N} T_{k=1}^{N} \left\{ I\left(T_{i=1}^{n} \left\{ \mu_{A_i^k}(\bar{x}_i) \right\}, \mu_{B^k}(\bar{y}^r) \right) \right\}}. \tag{6.5}$$

The specific form of formula (6.5) depends on the chosen definition of function I. Let us consider logical type systems which are constructed using definitions of triangular norms. Below we consider four cases of logical type neuro-fuzzy systems depending on the implication function:
(i) Neuro-fuzzy systems based on Łukasiewicz implication. In this case we obtain the following formula

$$\begin{aligned} \mu_{A^k \to B^k}(\bar{\mathbf{x}}, y) &= I\left(\mu_{A^k}(\bar{\mathbf{x}}), \mu_{B^k}(y)\right) = I\left(\mathop{T}_{k=1}^{n} \left(\mu_{A_i^k}(\bar{x}_i) \right), \mu_{B^k}(y) \right) \\ &= \min\left[1, 1 - \mathop{T}_{i=1}^{n} \left(\mu_{A_i^k}(\bar{x}_i) \right) + \mu_{B^k}(y) \right]. \end{aligned} \tag{6.6}$$

In view of (6.6) and (6.5) we get

$$\bar{y} = \frac{\sum_{r=1}^{N} \bar{y}^r T_{k=1}^{N} \left\{ \min\left[1, 1 - T_{i=1}^{n} \left(\mu_{A_i^k}(\bar{x}_i) \right) + \mu_{B^k}(\bar{y}^r) \right] \right\}}{\sum_{r=1}^{N} T_{k=1}^{N} \left\{ \min\left[1, 1 - T_{i=1}^{n} \left(\mu_{A_i^k}(\bar{x}_i) \right) + \mu_{B^k}(\bar{y}^r) \right] \right\}}. \tag{6.7}$$

and we obtain system description

$$\bar{y} = \frac{\sum_{r=1}^{N} \bar{y}^r T_{k=1}^{N} \left\{ \min\left[\begin{array}{l} 1, 1 - T_{i=1}^{n} \left(\exp\left[-\left(\frac{\bar{x}_i - \bar{x}_i^k}{\sigma_i^k} \right)^2 \right] \right) \\ + \exp\left[-\left(\frac{\bar{y}^r - \bar{y}^k}{\sigma^k} \right)^2 \right] \end{array} \right] \right\}}{\sum_{r=1}^{N} T_{k=1}^{N} \left\{ \min\left[\begin{array}{l} 1, 1 - T_{i=1}^{n} \left(\exp\left[-\left(\frac{\bar{x}_i - \bar{x}_i^k}{\sigma_i^k} \right)^2 \right] \right) \\ + \exp\left[-\left(\frac{\bar{y}^r - \bar{y}^k}{\sigma^k} \right)^2 \right] \end{array} \right] \right\}}. \tag{6.8}$$

(ii) Neuro-fuzzy systems based on binary implication. In this case we obtain the following formula

$$\begin{aligned} \mu_{A^k \to B^k}(\bar{\mathbf{x}}, y) &= I\left(\mu_{A^k}(\bar{\mathbf{x}}), \mu_{B^k}(y)\right) = I\left(\mathop{T}_{i=1}^{n} \left(\mu_{A_i^k}(\bar{x}_i) \right), \mu_{B^k}(y) \right) \\ &= \max\left[1 - \mathop{T}_{i=1}^{n} \left(\mu_{A_i^k}(\bar{x}_i) \right), \mu_{B^k}(y) \right]. \end{aligned} \tag{6.9}$$

In view of (6.9) and (6.5) we get

$$\bar{y} = \frac{\sum_{r=1}^{N} \bar{y}^r T_{k=1}^{N} \left\{ \max \left[1 - T_{i=1}^{n} \left(\mu_{A_i^k}(\bar{x}_i) \right), \mu_{B^k}(\bar{y}^r) \right] \right\}}{\sum_{r=1}^{N} T_{k=1}^{N} \left\{ \max \left[1 - T_{i=1}^{n} \left(\mu_{A_i^k}(\bar{x}_i) \right), \mu_{B^k}(\bar{y}^r) \right] \right\}}. \tag{6.10}$$

and finally we obtain the following fuzzy system description

$$\bar{y} = \frac{\sum_{r=1}^{N} \bar{y}^r T_{k=1}^{N} \left\{ \max \begin{bmatrix} 1 - T_{i=1}^{n} \left(\exp \left[- \left(\frac{\bar{x}_i - \bar{x}_i^k}{\sigma_i^k} \right)^2 \right] \right), \\ \exp \left[- \left(\frac{\bar{y}^r - \bar{y}^k}{\sigma^k} \right)^2 \right] \end{bmatrix} \right\}}{\sum_{r=1}^{N} T_{k=1}^{N} \left\{ \max \begin{bmatrix} 1 - T_{i=1}^{n} \left(\exp \left[- \left(\frac{\bar{x}_i - \bar{x}_i^k}{\sigma_i^k} \right)^2 \right] \right), \\ \exp \left[- \left(\frac{\bar{y}^r - \bar{y}^k}{\sigma^k} \right)^2 \right] \end{bmatrix} \right\}}. \tag{6.11}$$

(iii) Neuro-fuzzy systems based on Reichenbach implication. In this case we obtain the following formula

$$\begin{aligned} \mu_{A^k \to B^k}(\bar{\mathbf{x}}, y) &= I\left(\mu_{A^k}(\bar{\mathbf{x}}), \mu_{B^k}(y)\right) = I\left(\overset{n}{\underset{i=1}{T}} \left(\mu_{A_i^k}(\bar{x}_i) \right), \mu_{B^k}(y) \right) \\ &= 1 - \overset{n}{\underset{i=1}{T}} \left(\mu_{A_i^k}(\bar{x}_i) \right) (1 - \mu_{B^k}(y)). \end{aligned} \tag{6.12}$$

In view of (6.12) and (6.5) we get

$$\bar{y} = \frac{\sum_{r=1}^{N} \bar{y}^r T_{k=1}^{N} \left\{ 1 - T_{i=1}^{n} \left(\mu_{A_i^k}(\bar{x}_i) \right) (1 - \mu_{B^k}(\bar{y}^r)) \right\}}{\sum_{r=1}^{N} T_{k=1}^{N} \left\{ 1 - T_{i=1}^{n} \left(\mu_{A_i^k}(\bar{x}_i) \right) (1 - \mu_{B^k}(\bar{y}^r)) \right\}}. \tag{6.13}$$

and we obtain the neuro-fuzzy system description

$$\bar{y} = \frac{\sum_{r=1}^{N} \bar{y}^r T_{k=1}^{N} \left\{ \begin{array}{l} 1 - T_{i=1}^{n} \left(\exp \left[- \left(\frac{\bar{x}_i - \bar{x}_i^k}{\sigma_i^k} \right)^2 \right] \right) \\ \left(\left(1 - \exp \left[- \left(\frac{\bar{y}^r - \bar{y}^k}{\sigma^k} \right)^2 \right] \right) \right. \end{array} \right\}}{\sum_{r=1}^{N} T_{k=1}^{N} \left\{ \begin{array}{l} 1 - T_{i=1}^{n} \left(\exp \left[- \left(\frac{\bar{x}_i - \bar{x}_i^k}{\sigma_i^k} \right)^2 \right] \right) \\ \left(\left(1 - \exp \left[- \left(\frac{\bar{y}^r - \bar{y}^k}{\sigma^k} \right)^2 \right] \right) \right. \end{array} \right\}}. \tag{6.14}$$

(iv) Neuro-fuzzy systems based on Zadeh implication. In this case we obtain the following formula

$$\mu_{A^k \to B^k}(\overline{\mathbf{x}}, y) = I\left(\mu_{A^k}(\overline{\mathbf{x}}), \mu_{B^k}(y)\right) = I\left(\overset{n}{\underset{i=1}{T}}\left(\mu_{A_i^k}(\overline{x}_i)\right), \mu_{B^k}(y)\right) \tag{6.15}$$
$$= \max\left\{\min\left[\overset{n}{\underset{i=1}{T}}\left(\mu_{A_i^k}(\overline{x}_i)\right), \mu_{B^k}(y)\right], 1 - \overset{n}{\underset{i=1}{T}}\left(\mu_{A_i^k}(\overline{x}_i)\right)\right\}.$$

In view of (6.15) and (6.5) we get

$$\overline{y} = \frac{\sum_{r=1}^{N} \overline{y}^r T\left\{\begin{array}{l} \max\left\{\overset{n}{\underset{i=1}{T}}\left\{\mu_{A_i^r}(\overline{x}_i)\right\}, 1 - \overset{n}{\underset{i=1}{T}}\left\{\mu_{A_i^r}(\overline{x}_i)\right\}\right\}, \\ \overset{N}{\underset{\substack{k=1\\k\neq r}}{T}}\left\{\max\left\{\begin{array}{l}\min\left[\overset{n}{\underset{i=1}{T}}\left\{\mu_{A_i^k}(\overline{x}_i)\right\}, \mu_{B^k}(\overline{y}^r)\right], \\ 1 - \overset{n}{\underset{i=1}{T}}\left\{\mu_{A_i^k}(\overline{x}_i)\right\}\end{array}\right\}\right\}\end{array}\right\}}{\sum_{r=1}^{N} T\left\{\begin{array}{l} \max\left\{\overset{n}{\underset{i=1}{T}}\left\{\mu_{A_i^r}(\overline{x}_i)\right\}, 1 - \overset{n}{\underset{i=1}{T}}\left\{\mu_{A_i^r}(\overline{x}_i)\right\}\right\}, \\ \overset{N}{\underset{\substack{k=1\\k\neq r}}{T}}\left\{\max\left\{\begin{array}{l}\min\left[\overset{n}{\underset{i=1}{T}}\left\{\mu_{A_i^k}(\overline{x}_i)\right\}, \mu_{B^k}(\overline{y}^r)\right], \\ 1 - \overset{n}{\underset{i=1}{T}}\left\{\mu_{A_i^k}(\overline{x}_i)\right\}\end{array}\right\}\right\}\end{array}\right\}}. \tag{6.16}$$

and we obtain the neuro-fuzzy system description based on Zadeh implication

$$\overline{y} = \frac{\sum_{r=1}^{N} \overline{y}^r T\left\{\begin{array}{l} \max\left\{\begin{array}{l}\overset{n}{\underset{i=1}{T}}\left\{\exp\left[-\left(\frac{\overline{x}_i - \overline{x}_i^r}{\sigma_i^r}\right)^2\right]\right\}, \\ 1 - \overset{n}{\underset{i=1}{T}}\left\{\exp\left[-\left(\frac{\overline{x}_i - \overline{x}_i^r}{\sigma_i^r}\right)^2\right]\right\}\end{array}\right\}, \\ \overset{N}{\underset{\substack{k=1\\k\neq r}}{T}}\left\{\max\left\{\begin{array}{l}\min\left\{\begin{array}{l}\overset{n}{\underset{i=1}{T}}\left\{\exp\left[-\left(\frac{\overline{x}_i - \overline{x}_i^k}{\sigma_i^k}\right)^2\right]\right\}, \\ \exp\left[-\left(\frac{\overline{y}^r - \overline{y}^k}{\sigma^k}\right)^2\right]\end{array}\right\}, \\ 1 - \overset{n}{\underset{i=1}{T}}\left\{\exp\left[-\left(\frac{\overline{x}_i - \overline{x}_i^k}{\sigma_i^k}\right)^2\right]\right\}\end{array}\right\}\right\}\end{array}\right\}}{\sum_{r=1}^{N} T\left\{\begin{array}{l} \max\left\{\begin{array}{l}\overset{n}{\underset{i=1}{T}}\left\{\exp\left[-\left(\frac{\overline{x}_i - \overline{x}_i^r}{\sigma_i^r}\right)^2\right]\right\}, \\ 1 - \overset{n}{\underset{i=1}{T}}\left\{\exp\left[-\left(\frac{\overline{x}_i - \overline{x}_i^r}{\sigma_i^r}\right)^2\right]\right\}\end{array}\right\}, \\ \overset{N}{\underset{\substack{k=1\\k\neq r}}{T}}\left\{\max\left\{\begin{array}{l}\min\left\{\begin{array}{l}\overset{n}{\underset{i=1}{T}}\left\{\exp\left[-\left(\frac{\overline{x}_i - \overline{x}_i^k}{\sigma_i^k}\right)^2\right]\right\}, \\ \exp\left[-\left(\frac{\overline{y}^r - \overline{y}^k}{\sigma^k}\right)^2\right]\end{array}\right\}, \\ 1 - \overset{n}{\underset{i=1}{T}}\left\{\exp\left[-\left(\frac{\overline{x}_i - \overline{x}_i^k}{\sigma_i^k}\right)^2\right]\right\}\end{array}\right\}\right\}\end{array}\right\}}. \tag{6.17}$$

Further in this chapter we will use the so called simplified logical fuzzy systems. In such systems output fuzzy sets B^k are only marginally overlapped or totally separated from each other. In this situation, the condition $\mu_{B^k}(\overline{y}^r) \approx 0$ is satisfied.

Above cases (i)–(III) belongs to one group of S–implications. If $\mu_{B^k}(\overline{y}^r) \approx 0$, then neuro-fuzzy system descriptions (6.8), (6.11) and (6.14) will be reduced to the following, simplified form

$$\overline{y} = \frac{\sum_{r=1}^{N} \overline{y}^r \mathop{T}_{\substack{k=1 \\ k\neq r}}^{N} \left\{ 1 - \mathop{T}_{i=1}^{n} \left\{ \exp\left[-\left(\frac{\overline{x}_i - \overline{x}_i^k}{\sigma_i^k} \right)^2 \right] \right\} \right\}}{\sum_{r=1}^{N} \mathop{T}_{\substack{k=1 \\ k\neq r}}^{N} \left\{ 1 - \mathop{T}_{i=1}^{n} \left\{ \exp\left[-\left(\frac{\overline{x}_i - \overline{x}_i^k}{\sigma_i^k} \right)^2 \right] \right\} \right\}}. \tag{6.18}$$

Similarly, if $\mu_{B^k}(\overline{y}^r) \approx 0$, then we will obtain a simplified Zadeh structure given by the formula

$$\overline{y} = \frac{\sum_{r=1}^{N} \overline{y}^r T \left\{ \begin{array}{l} \max \left\{ \begin{array}{l} \mathop{T}_{i=1}^{n} \left\{ \exp\left[-\left(\frac{\overline{x}_i - \overline{x}_i^r}{\sigma_i^r} \right)^2 \right] \right\}, \\ 1 - \mathop{T}_{i=1}^{n} \left\{ \exp\left[-\left(\frac{\overline{x}_i - \overline{x}_i^r}{\sigma_i^r} \right)^2 \right] \right\} \end{array} \right\}, \\ \mathop{T}_{\substack{k=1 \\ k\neq r}}^{N} \left\{ 1 - \mathop{T}_{i=1}^{n} \left\{ \exp\left[-\left(\frac{\overline{x}_i - \overline{x}_i^k}{\sigma_i^k} \right)^2 \right] \right\} \right\} \end{array} \right\}}{\sum_{r=1}^{N} T \left\{ \begin{array}{l} \max \left\{ \begin{array}{l} \mathop{T}_{i=1}^{n} \left\{ \exp\left[-\left(\frac{\overline{x}_i - \overline{x}_i^r}{\sigma_i^r} \right)^2 \right] \right\}, \\ 1 - \mathop{T}_{i=1}^{n} \left\{ \exp\left[-\left(\frac{\overline{x}_i - \overline{x}_i^r}{\sigma_i^r} \right)^2 \right] \right\} \end{array} \right\}, \\ \mathop{T}_{\substack{k=1 \\ k\neq r}}^{N} \left\{ 1 - \mathop{T}_{i=1}^{n} \left\{ \exp\left[-\left(\frac{\overline{x}_i - \overline{x}_i^k}{\sigma_i^k} \right)^2 \right] \right\} \right\} \end{array} \right\}}. \tag{6.19}$$

In the above systems the following parameters of the membership functions are subject to learning: $\overline{x}_i^k, \sigma_i^k, \overline{y}^k, \sigma^k$. In simplified systems, the parameters $\overline{x}_i^k, \sigma_i^k, \overline{y}^k$ are subject to learning. Structure described by (6.18) is shown in Fig. 6.1.

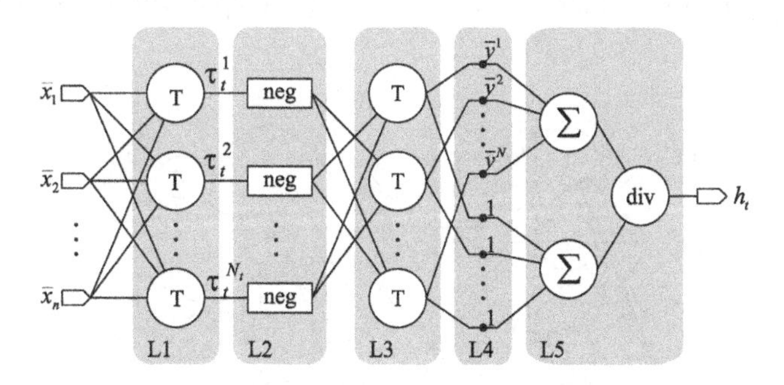

Fig. 6.1 General structure of logical neuro-fuzzy system based on S-implication

6.2 Interpretable Boosting Ensembles of Logical Models

This section describes the method to build an ensemble of logical type neuro-fuzzy systems with the possibility to merge all rule bases. We consider systems based on s-implications (6.18). To build the ensemble we use the AdaBoost algorithm described in section 3.2. As already mentioned in section 4.3, the most important problem in the case of creating ensembles from fuzzy systems as base hypothesis is that each rule base has different sum of rule activation levels. Thus we cannot treat them as one large fuzzy rule base. Later on in this section simplified neuro-fuzzy system based on S-implication, see (6.18), will be used. After putting (6.18) into the formula for the AdaBoost ensemble (3.9), we obtain the final output of the classifier ensemble

$$f(\mathbf{x}) = \sum_{t=1}^{T} c_t \frac{\sum\limits_{r=1}^{N_t} \bar{y}_t^r \cdot \mathop{T}\limits_{\substack{k=1 \\ k \neq r}}^{N_t} \left[1 - \tau_t^k\right]}{\sum\limits_{r=1}^{N_t} \mathop{T}\limits_{\substack{k=1 \\ k \neq r}}^{N_t} \left[1 - \tau_t^k\right]} \tag{6.20}$$

where Formula (6.20) is the weighted sum of all hypothesis outcomes. It is not possible to merge all rule bases coming from members of the ensemble, because (6.20) cannot be rewritten to use a common denominator. Let us observe that in the denominator of (6.20) there is the sum of activity levels of rules in a single neuro-fuzzy system. Thus if we want to treat formula (6.20) as one neuro-fuzzy system, it should be transformed so that the sum in the denominator equals 1. Considered neuro-fuzzy systems should be trained to satisfy two assumptions

$$\begin{array}{l} 1)\ h_t(\mathbf{x}^l) = d^l \ \forall l = 1,...,m \\ 2)\ \sum\limits_{r=1}^{N_t} \mathop{T}\limits_{\substack{k=1 \\ k \neq r}}^{N_t} \left[1 - \tau_t^k\right] = 1 \ \forall t = 1,...,T \end{array} \tag{6.21}$$

To satisfy (6.21) we transform the neuro-fuzzy structure in Fig. 6.1 to obtain the form depicted in Fig. 6.2. We removed the last layer performing the division, thus the system has two outputs. The error on the first output will be computed taking

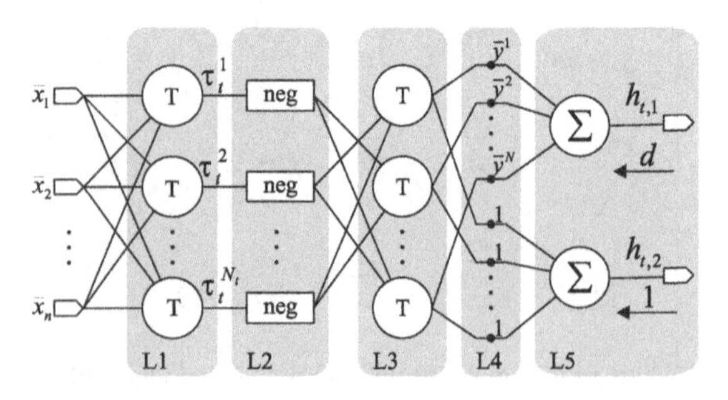

Fig. 6.2 Single modified logical neuro-fuzzy system

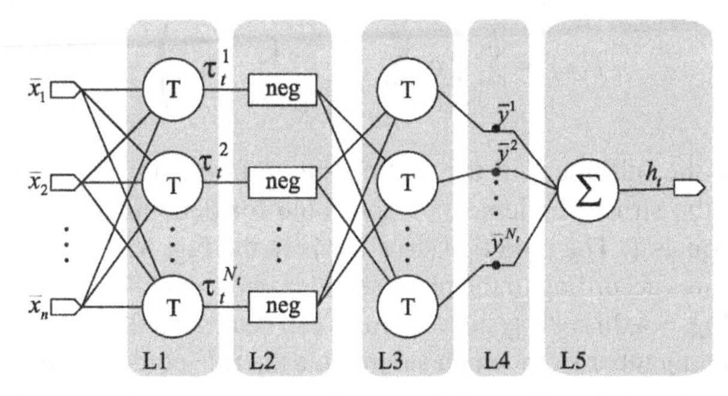

Fig. 6.3 Single logical neuro-fuzzy system after learning. The second output is removed as it always equals one.

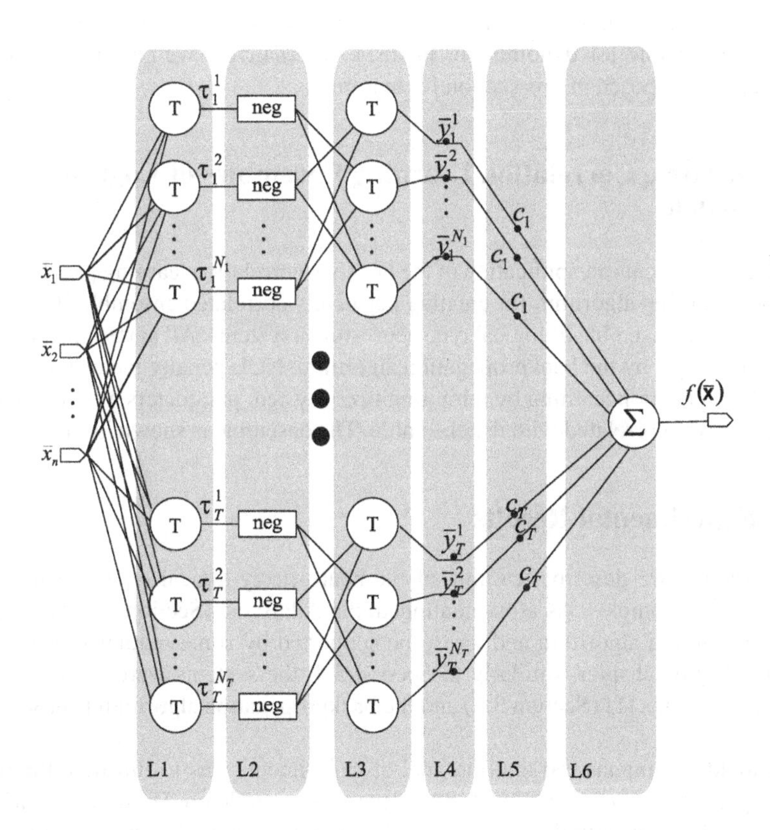

Fig. 6.4 Ensemble of modified logical neuro-fuzzy systems combined after learning.

into account desired output from the learning data. Desired signal on the second output is constant and equals 1. If (6.21) holds then the following formula is valid for the output of the whole ensemble

$$f(\mathbf{x}) = \sum_{t=1}^{T} \left(c_t \sum_{r=1}^{N_t} \bar{y}_t^r \cdot \underset{\substack{k=1 \\ k \neq r}}{\overset{N_t}{T}} \left[1 - \tau_t^k \right] \right) \tag{6.22}$$

As we assume the fulfilment of the first condition in (6.21), we can remove after learning from the structure elements responsible for denominator in (6.1), as the denominator equals 1. The modified structure is in the Fig. 6.3. After learning consecutive structures according to the proposed idea, we can build the modular system presenting in Fig. 6.4. In such system the rules can be arranged in arbitrary order. We do not have to remember from which submodule the rule comes from. The obtained ensemble of neuro-fuzzy systems can be interpreted as a regular single neuro-fuzzy system, where coefficients c_t can be interpreted as weights of fuzzy rules. Important ability of the system is its possible fine tuning. Yet, in order to make the structure acting like a neuro-fuzzy system during learning, we have to add the removed elements responsible for denominator in (6.1). Alternatively we can use again the modification proposed in this section for learning.

6.3 Negative Correlation Learning Ensembles of Logical Models

This section presents an application of the negative correlation learning [3, 4], which is a meta-learning algorithm for creating negatively correlated ensembles (see Section 3.3), to an ensemble of logical type neuro-fuzzy systems. All system parameters were determined by the backpropagation algorithm. NCL penalty factor influenced the backpropagation learning by inhibiting fuzzy system parameters modification if the system was correlated with the ensemble. The ensemble is shown in Figure 6.5.

6.4 Experimental Results

In this section we describe learning of the neuro-fuzzy logical type systems and experimental examples. As aforementioned, the AdaBoost (Section 3.2) algorithm is a metalearning algorithm and has to be supported by some regular learning algorithm. In this chapter, similarly to Section 5.4, the systems were initialized by the fuzzy c-means [1] (Section 3.5) and the backpropagation algorithm [7] (Section 3.4).

It should be emphasized that the goal of experiments was to illustrate the proposed approach and to show the ability of the systems to learn. We do not compare the obtained results with the results from the literature, as the aim of the paper was to show the possibility to merge all systems from the ensemble into one rule base.

Logical type neuro-fuzzy system knowledge is expressed in the form of fuzzy rules. During learning, fuzzy set parameters have to be determined. At the beginning, antecedent fuzzy sets are determined by the fuzzy c-means clustering algorithm. Then, all parameters, i.e. antecedent and consequent fuzzy sets parameters, are tuned by the backpropagation algorithm [7]. Having given learning data set of

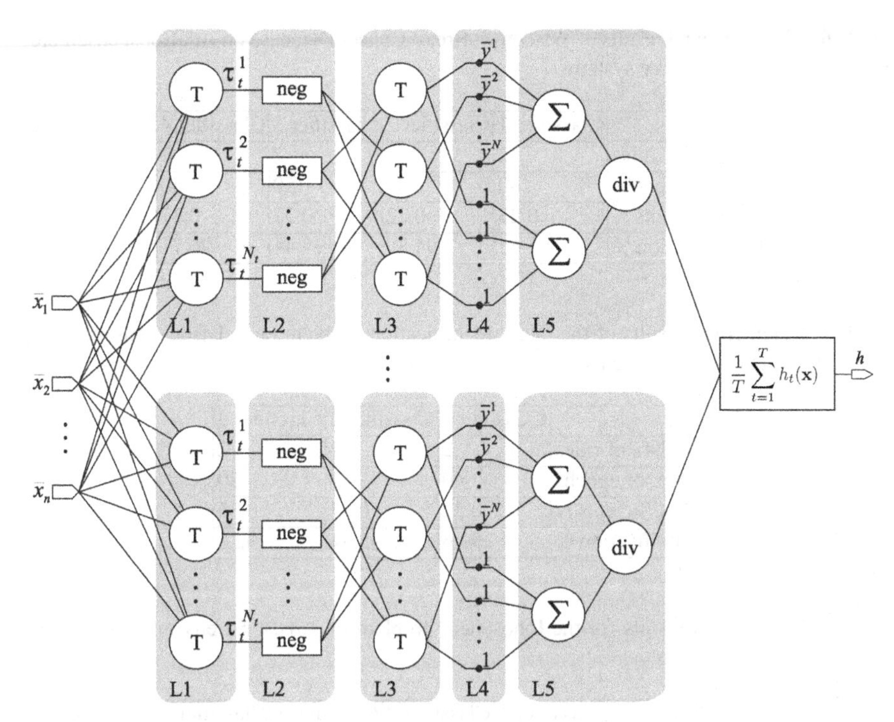

Fig. 6.5 Negative correlation learning ensemble of logical neuro-fuzzy systems.

pair $(\bar{\mathrm{x}}, d)$ where d is the desired response of the system we can use the following error measure

$$Q(\bar{\mathbf{x}}, d) = \tfrac{1}{2}\left[\bar{y}(\bar{\mathbf{x}}) - d\right]^2 \tag{6.23}$$

Every neuro-fuzzy system parameter, denoted for simplicity as p, can be determined by minimizing the error measure in the iterative procedure. For every iteration t, the parameter value is computed by

$$p(t+1) = p(t) - \eta \frac{\partial Q(\bar{\mathbf{x}}, d; t)}{\partial p(t)} \tag{6.24}$$

where η is a learning coefficient, set in simulations to 0.03. Datasets used in experiments were divided randomly into a learning set and a testing set.

The first experiment was carried out on the Wisconsin Breast Cancer Database [2] (Section 1.2.4). From the data set 205 instances were taken into testing data and 16 instances with missing features were removed. The classification accuracy was 96.7%. Detailed errors and subsystem parameters are described in Table 6.1. The second experiment was performed on the Glass Identification dataset the Glass Identification problem [2] (see Section 1.2.1). We took out 43 instances for a testing set. We obtained 95.34% classification accuracy. We used three logical type neuro-fuzzy classifiers described in Table 6.2. Simulations shows superior performance

Table 6.1 Experimental results of Wisconsin Breast Cancer Database on AdaBoost ensemble of logical type neuro-fuzzy systems

	Classifier 1	Classifier 2	Classifier 3	Classifier 4
No. of rules	2	2	2	2
No. of epochs	40	35	50	40
Coefficients c_t	0.23	0.22	0.27	0.27
Classification %	95.62	94.89	97.44	96.71

Table 6.2 Numerical results of the Glass Identification problem on AdaBoost ensemble of logical type neuro-fuzzy systems

	Classifier 1	Classifier 2	Classifier 3
No. of rules	2	2	2
No. of epochs	40	37	50
Coefficients c_t	0.32	0.33	0.35
Classification %	97.6	95.34	92.8

Table 6.3 Numerical results for the Ionosphere problem on AdaBoost ensemble of logical type neuro-fuzzy systems

	Classifier 1	Classifier 2	Classifier 3	Classifier 4
No. of rules	2	2	2	2
No. of epochs	35	35	35	50
Coefficients c_t	0.25	0.23	0.25	0.27
Classification %	94.34	94.31	93.54	92.23

and faster convergence from the proposed approach comparing to those from the literature. The next problem used in the simulations was the Ionosphere problem [2] (see Section 1.2.2). The data were divided randomly into 246 learning and 105 testing instances. We used 4 logical type neuro-fuzzy classifiers described in Table 6.3. We obtained 93.62% classification accuracy.

Apart from the aforementioned AdaBoost ensembles, we simulated also the NCL ensemble of logical type systems (Section 3.3). The systems were initialized randomly by the fuzzy c-means clustering and then trained the backpropagation algorithm. From the WBCD data set 205 instances were taken into testing data and 16 instances with missing features were removed. The classification accuracy was 95.4%. In case of the Glass Identification problem we took out 43 instances for a testing set. We obtained 94.26% classification accuracy. We used 3 neuro-fuzzy classifiers. In the Ionosphere problem [2] (see Section 1.2.2) the data were divided randomly into 246 learning and 105 testing instances. We used 5 neuro-fuzzy classifiers, we obtained 94.12% classification accuracy.

6.5 Summary and Discussion

This chapter shows the method to join fuzzy rule bases from several ensembled logical type fuzzy systems into one rule base. The idea of logical type neuro-fuzzy systems is described as they were chosen for members of the ensemble. The reason for this choice was a better suitability of the logical type fuzzy systems for classification tasks over the common Mamdani methods. The ensemble was designed on the basis of the AdaBoost meta-learning. The hypothesis within the ensemble were trained by the fuzzy c-means algorithm and the backpropagation algorithm. Thanks to inherent interpretability in fuzzy systems, it was easy to initialize the system using some expert knowledge or exploratory data analysis methods. The essential problem in the case of fuzzy system ensembles is solved - fuzzy rules coming from different fuzzy systems cannot be merged immediately. We proposed a method to overcome this disadvantage. The fuzzy classifiers are modified in a specific way to keep the activation level of each rule base at the same level. Thus none of classifiers dominates over the ensemble. The modification is very simple and does not introduce additional computational burden. The fundamental results of this chapter were partly presented in [6].

In the experiments, fuzzy set membership functions are set in advance by the fuzzy c-means algorithm. Then all system parameters are determined by the backpropagation algorithm. Boosting weights influence the backpropagation learning by inhibiting fuzzy system parameter modification for a given sample. Simulations on some popular benchmarks show great accuracy of the proposed methods. However most important advantage of the proposed method is the possibility of merging several separate fuzzy rule bases into one set of fuzzy rules.

References

1. Bezdek, J., Keller, J., Krisnapuram, R., Pal, N.: Fuzzy Models and Algorithms for Pattern Recognition and Image Processing. Kluwer Academic Press (1999)
2. Frank, A., Asuncion, A.: UCI machine learning repository (2010), `http://archive.ics.uci.edu/ml`
3. Liu, Y., Yao, X.: Ensemble learning via negative correlation. Neural Networks 12, 1399–1404 (1999)
4. Liu, Y., Yao, X.: Simultaneous training of negatively correlated neural networks in an ensemble. IEEE Trans. Syst., Man, Cybern. B 29, 716–725 (1999)
5. Rutkowski, L.: Flexible Neuro-Fuzzy Systems. Kluwer Academic Publishers (2004)
6. Scherer, R.: An ensemble of logical-type neuro-fuzzy systems. In: Expert Systems With Applications (2011), doi:10.1016/j.eswa.2011.04.117
7. Wang, L.X.: Adaptive Fuzzy Systems and Control. PTR Prentice-Hall, Englewood Cliffs (1994)

Chapter 7
Takagi-Sugeno Fuzzy Systems

The Takagi-Sugeno systems (for short, to be denoted TS) are one of the most common fuzzy models. In such systems consequents are functions of inputs. This chapter shows a modification of such models as members of an classifier ensemble. The problem of incapability of merging several rule bases is addressed by a novel design of fuzzy systems constituting the ensemble, resulting in normalization of individual rule bases during learning.

7.1 Takagi-Sugeno Systems

We use the Takagi-Sugeno fuzzy systems with linear consequents as classifiers in the ensemble. In general case, fuzzy rules in TS systems take the following form

$$\begin{array}{l} R^r\colon \mathrm{IF} x_1 \mathrm{is} A_1^r \mathrm{AND} \ldots \mathrm{AND} x_n \mathrm{is} A_n^r \\ \mathrm{THEN} y = f^{(r)}\left(x_1, x_2, \ldots, x_n\right) \end{array} \tag{7.1}$$

where $r = 1, \ldots, R$ is the rule number. The inputs are denoted by x_i, $i = 1, \ldots, n$. The function used in our model is linear, i.e.

$$\begin{array}{l} R^r : \mathrm{IF} x_1 \mathrm{is} A_i^r \mathrm{AND} \ldots \mathrm{AND} x_n \mathrm{is} A_n^r \\ \mathrm{THEN} y^r = a_0^{(r)} + a_1^{(r)} x_1 +, \ldots, + a_n^{(r)} x_n \end{array} \tag{7.2}$$

We use antecedent fuzzy sets which are represented by Gaussian functions

$$\mu_{A_i^r}(x_i) = \exp\left(-\left(\frac{x_i - \bar{x}_i^r}{\sigma_i^r}\right)^2\right) \tag{7.3}$$

where σ_i^r and $\bar{x}_i^r$ are, respectively, width and center of the function for the fuzzy set in the r-th rule and the i-th input. After presenting input signal $\bar{\mathbf{x}} = [\bar{x}_1, ..., \bar{x}_n]$, we compute the output of the Takagi-Sugeno system as the mean of y^r, $r = 1, \ldots, N$, weighted by rule activation levels

R. Scherer: Multiple Fuzzy Classification Systems, STUDFUZZ 288, pp. 73–79.
springerlink.com

$$\bar{y} = \frac{\sum_{r=1}^{N} y^r \overset{n}{\underset{i=1}{T}} \left(\mu_{A_i^r}(\bar{x}_i) \right)}{\sum_{r=1}^{N} \overset{n}{\underset{i=1}{T}} \left(\mu_{A_i^r}(\bar{x}_i) \right)} \tag{7.4}$$

Rule activation level of the k-th rule, denoted by τ^r, is defined by

$$\tau^r = \prod_{i=1}^{n} \mu_i^r(x_i) \tag{7.5}$$

We use systems (7.4) as members of an ensemble designed with the AdaBoost algorithm (see Section 3.2) which is the most popular boosting method [4, 5]. To compute the overall output of the ensemble of the Takagi-Sugeno fuzzy classifiers trained by AdaBoost algorithm the following formula is used

$$f(\mathbf{x}) = \sum_{t=1}^{T} c_t h_t(\mathbf{x}) , \tag{7.6}$$

where

$$c_t = \frac{\alpha_t}{\sum_{t=1}^{T} \alpha_t} \tag{7.7}$$

is classifier importance for a given training set, $h_t(\mathbf{x})$ is the response of the hypothesis t on the basis of feature vector $\mathbf{x} = [x_1, ..., x_n]$. The coefficient c_t value is computed on the basis of the classifier error and can be interpreted as the measure of classification accuracy of the given classifier. Moreover the assumption (7.9) should be met. As we see, the AdaBoost algorithm is a meta-learning algorithm and does not determine the way of learning for classifiers in the ensemble. To learn the ensemble members we can use any algorithm which is able to take into account learning vectors weights D_t^l. We train the structures by the backpropagation algorithm considering boosting learning sample weights D_t^l. The output of the whole ensemble consisting of the systems (7.4) takes the following form

$$h_t = \sum_{t=1}^{T} c_t \frac{\sum_{r=1}^{N_t} y_t^r \tau_t^r}{\sum_{r=1}^{N} \tau_t^r} \tag{7.8}$$

where $t = 1, ..., T$ is classifier number, and $r = 1, ..., N_t$ is the rule number. Takagi-Sugeno system, number t, is presented in Fig. 7.1.

7.2 Normalization of Rule Bases within Ensemble

Analogously to previous chapters we will show the method to merge all fuzzy rule bases from one boosting ensemble of the Takagi–Sugeno systems. The most important problem in the case of creating ensembles from fuzzy systems as base

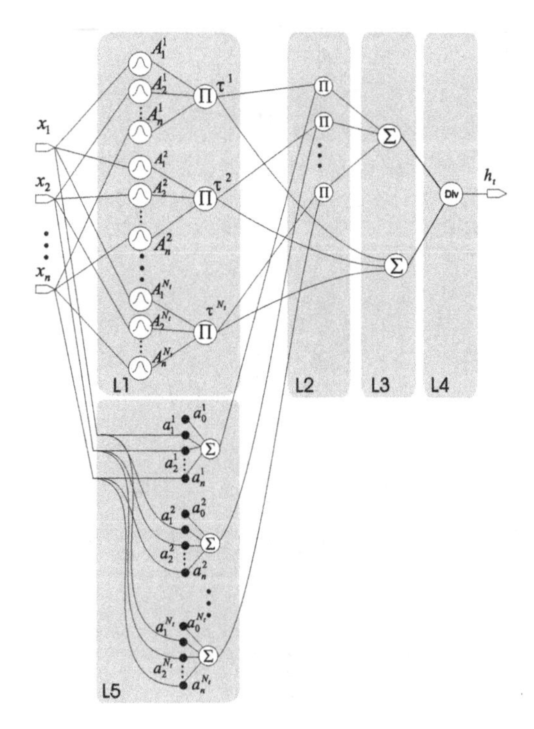

Fig. 7.1 Single Takagi-Sugeno system

hypothesis is that each rule base has different sum of rule activation levels. The normalization in this chapter is achieved by adding the second output to the system and keeping all rule bases at the same level. The structure of the Takagi–Sugeno systems will be modified to keep rule bases of all systems at same activity level (firing strength). The modification will consist in adding the second, normalizing output. The final output of the classifier ensemble is expressed by formula (7.8) which is the weighted sum of all hypothesis outcomes. Let us observe that in the denominator of (7.8) there is the sum of activity levels of rules in a single neuro-fuzzy system. Thus if we want to treat formula (7.8) as one neuro-fuzzy system, it should be transformed so that the sum in the denominator equals 1. Considered neuro-fuzzy systems should be trained to satisfy two assumptions

$$\begin{aligned} &1)\ h_t(\mathbf{x}^q) = d^q \ \forall q = 1,...,M \\ &2)\ \sum_{r=1}^{N_t} \tau_t^r = 1 \ \forall t = 1,...,T \end{aligned} \tag{7.9}$$

To satisfy (7.9) we transform the neuro-fuzzy structure in Fig. 7.1 to obtain the form depicted in Fig. 7.2. We removed the last layer performing the division, thus the system has two outputs. The error on the first output will be computed taking into account desired output from the learning data. Desired signal on the second

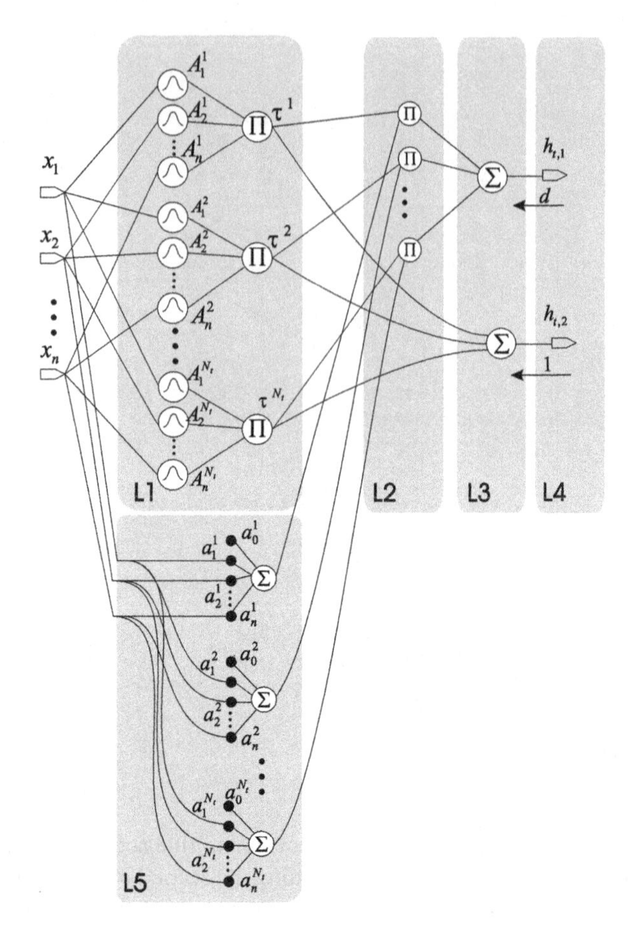

Fig. 7.2 Single Takagi-Sugeno system modified to normalize activation levels of all systems in the boosting ensemble

output is constant and equals 1. If (7.9) holds then the following formula is valid for the output of the whole ensemble

$$f(\mathbf{x}) = \sum_{t=1}^{T} \left(c_t \sum_{r=1}^{N_t} \bar{y}_t^r \cdot \tau_t^r \right) \tag{7.10}$$

The assumption (7.9) guarantees that in the case of creating an ensemble of several such modified Takagi-Sugeno neuro-fuzzy systems, activity level of one system does not dominate other subsystems. Since we assume the fulfilment of the first condition in (7.9), after learning we can remove elements responsible for denominator in (7.8) from the structure, as the denominator equals 1. The modified structure is shown in Fig. 7.3. After learning consecutive structures according to the proposed idea, we can build the modular system consisted of systems presented in Fig. 7.3.

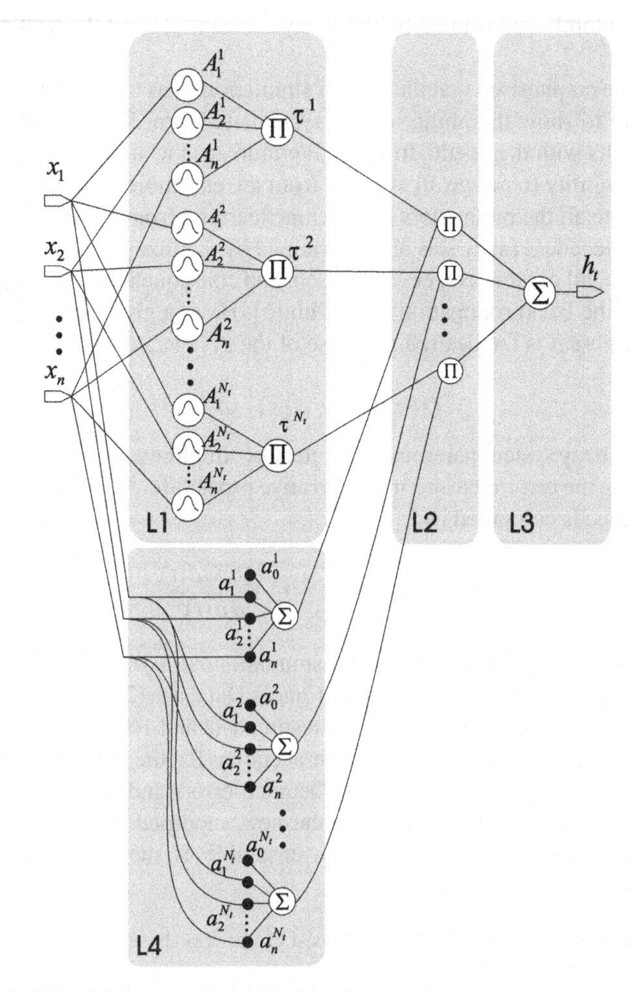

Fig. 7.3 Single Takagi-Sugeno system after learning process

In such a system the rules can be arranged in an arbitrary order. We do not have to remember from which submodule the rule comes from. The obtained ensemble of neuro-fuzzy systems can be interpreted as a regular single neuro-fuzzy system, where coefficients c_t can be interpreted as weights of fuzzy rules. Important ability of the merged system is its possible fine tuning.

7.3 Experimental Results

In this section we describe learning of the Takagi-Sugeno systems and experimental examples. As aforementioned, the AdaBoost algorithm is a metalearning algorithm and has to be supported by some regular learning algorithm. The systems were initialised randomly and trained by the fuzzy c-means [1] and the backpropagation algorithm [6].

It should be emphasised that the goal of simulations was to illustrate the proposed approach and to show the ability of the systems to learn. We do not compare the obtained results with the results from the literature, as the aim of the chapter was to show the possibility to merge all systems from the ensemble into one rule base.

We compute all the parameters by machine learning from numerical data. At the beginning, antecedent fuzzy sets are determined by the fuzzy c-means clustering algorithm. Then, all parameters, i.e. antecedent and consequent fuzzy sets parameters, are tuned by the backpropagation algorithm [6]. Having given learning data set of pair $(\bar{x}, d)$, where d is the desired response of the system, we can use the following error measure

$$Q(\bar{\mathbf{x}}, d) = \frac{1}{2}\left[\bar{y}(\bar{\mathbf{x}}) - d\right]^2 \tag{7.11}$$

Every neuro-fuzzy system parameter, denoted for simplicity as p, can be determined by minimizing the error measure in the iterative procedure. For every iteration t, the parameter value is computed by

$$p(t+1) = p(t) - \eta \frac{\partial Q(\bar{\mathbf{x}}, d; t)}{\partial p(t)} \tag{7.12}$$

where η is a learning coefficient, set in simulations to 0.02. The first simulations were carried out on the Wisconsin Breast Cancer Database [2] (Section 1.2.4). From the data set 205 instances were taken into testing data and 16 instances with missing features were removed. The classification accuracy for the ensemble of the modified Takagi-Sugeno system was 97.8%. Detailed errors and subsystem parameters are described in Table 7.1. To compare the new, modified systems with the standard Takagi-Sugeno ones, we learned an ensemble of such unmodified systems.

Table 7.1 Numerical results of Wisconsin Breast Cancer Database

	Classif. 1	Classif. 2	Classif. 3	Classif. 4
No. of rules	2	2	2	2
No. of epochs	30	30	50	70
Coefficients c_t	0.24	0.23	0.27	0.27
Classification %	88.37	76.74	90.6	97.67

Table 7.2 Numerical results of Wisconsin Breast Cancer Database for the unmodified Takagi-Sugeno systems

	Classif. 1	Classif. 2	Classif. 3	Classif. 4
No. of rules	2	2	2	2
No. of epochs	40	30	30	60
Coefficients c_t	0.45	0.17	0.19	0.19
Classification %	98.54	76.74	97.80	97.80

The obtained accuracy was slightly higher - 99.6% and detailed system parameters are presented in Tab. 7.2.

7.4 Summary and Discussion

Classifier ensembles are popular method for increasing the accuracy. This chapter deals with the ensemble of the Takagi-Sugeno systems. When we create such an ensemble, after learning we obtain several fuzzy rule bases. Such fuzzy rules, coming from different fuzzy systems, cannot be merged immediately. We proposed a method to overcome this disadvantage. The fuzzy classifiers are modified to normalize consecutive rule bases. The fundamental results of this chapter were partly presented in [3]. In the numerical simulations, fuzzy set membership functions are set in advance by the fuzzy c-means algorithm. Then all system parameters are determined by the backpropagation algorithm. Boosting weights influence the backpropagation learning by inhibiting fuzzy system parameter modification for a given sample. Simulations show good accuracy of the proposed method. However most important advantage of the proposed method is the possibility of merging several separate fuzzy rule bases into one set of fuzzy rules. The ensemble of unmodified systems performs somewhat better but in this case we do have the possibility to merge all fuzzy rules and potentially to work farther toward interpretable fuzzy systems what is impossible in the case of the unmodified systems.

References

1. Bezdek, J., Keller, J., Krisnapuram, R., Pal, N.: Fuzzy Models and Algorithms for Pattern Recognition and Image Processing. Kluwer Academic Press (1999)
2. Frank, A., Asuncion, A.: UCI machine learning repository (2010), `http://archive.ics.uci.edu/ml`
3. Korytkowski, M., Rutkowski, L., Scherer, R.: Rule base normalization in takagi-sugeno ensemble. In: IEEE Symposium Series on Computational Intelligence - SSCI 2011, Paris, France, 2011 IEEE Workshop on Hybrid Intelligent Models and Applications, April 11-15, pp. 1–5 (2011)
4. Meir, R., Rätsch, G.: An Introduction to Boosting and Leveraging. In: Mendelson, S., Smola, A.J. (eds.) Advanced Lectures on Machine Learning. LNCS (LNAI), vol. 2600, pp. 118–183. Springer, Heidelberg (2003)
5. Schapire, R.E.: A brief introduction to boosting. In: Conference on Artificial Intelligence, pp. 1401–1406 (1999)
6. Wang, L.X.: Adaptive Fuzzy Systems and Control. PTR Prentice-Hall, Englewood Cliffs (1994)

Chapter 8
Rough–neuro–fuzzy Ensembles for Classification with Missing Data

Neuro-fuzzy systems presented so far in the book are not able to cope with missing data. Generally, there are two ways to solve the problem of missing data:

- *Imputation* - the unknown values are replaced by estimated ones [2, 4, 23, 30]. The estimated value can be set as the mean of known values of the same feature in other instances. An another idea is to apply the nearest neighbor algorithm based on instances with known value of the same feature [13]. The statistical method can be also used [1, 19].
- *Marginalization* - the features with unknown values are ignored [3]. In this way the problem comes down to the classification in lower dimensional feature space.

This chapter presents a new approach to ensemble fuzzy classification in the case of missing data. The rough fuzzy sets are incorporated into various Mamdani and logical type neuro-fuzzy structures and the rough-neuro-fuzzy classifier is derived. Theorems which allow to determine the structure of a rough-neuro-fuzzy classifier are given. Several experiments illustrating the performance of the rough-neuro-fuzzy ensemble are described. The experiments were performed with gradually increasing the number of missing features.

8.1 Basic Definitions of the Rough Set Theory

This chapter presents the rough set theory [21] combined with fuzzy classifier ensemble techniques as a solution to the problem of missing data. The object can be classified to the positive region of the class (i.e. object certainly belongs to the class), to the negative region of the class (i.e. object certainly not belongs to the class) or to the boundary region of the class (i.e. it is not possible to determine if object belongs to the class or not). The more detailed description of the object (more features available), the more detailed classification accuracy.

First, we shall define the universe of discourse U. It is the set of all objects which constitute the area of our interest. A single j-th element of this space will be denoted as x_j. Each object of the space U may be characterized using specific features. If it is a physical object, most certainly it has infinitely many features, however, we shall

R. Scherer: Multiple Fuzzy Classification Systems, STUDFUZZ 288, pp. 81–127.
springerlink.com

limit the selection to their specific subset. Let us denote the interesting set of object features of space U by the symbol Q. Let us denote the individual features by the symbol q appropriately indexed, e.g. q_i. What differentiates one object from another and makes other objects similar, these are the values of their features. Let us denote by V_q the set of values that the feature q can take. The value of feature q of the object x will be denoted as v_q^x. The vector of all object x features may be denoted as $\mathbf{v}^x = \left[v_{q1}^x, v_{q2}^x, ..., v_{qn}^x\right]$. One of methods to present the information about objects characterized by the same set of features is the information system.

Definition 8.1. *The information system* is referred to the ordered 4-tuple $SI = \langle U, Q, V, f \rangle$ [1], where U is the set of objects, Q is the set of features (attributes), $V = \bigcup_{q \in Q} V_q$ is the set of all possible values of features, while $f : U \times Q \rightarrow V$ is called the information function. We can say that $v_q^x = f(x, q)$, of course $f(x, q) \in V_q$. The notation $v_q^x = f_x(q)$, which treats the information function as a family of functions, will be considered as equivalent. Then $f_x : Q \rightarrow V$.

The special case of the information system is the decision table.

Definition 8.2. *The decision table* is the ordered 5-tuple $DT = \langle U, C, D, V, f \rangle$. The elements of the set C we call conditional features (attributes), and elements of D-decision features (attributes).

The information function f described in Definition 8.1 defines unambiguously the set of rules included in the decision table. In the notation, in the form of family of functions, the function $f_l : C \times D \rightarrow V$ defines l th decision rule of the table. The difference between the above definition and Definition 8.1 consists in separation of the set of features Q into two disjoint subsets C and D, complementary to Q. The decision tables are an alternative way of representing the information with relation to the rules:

$$R^l : \textbf{IF } c_1 = v_{c_1}^l \textbf{ AND } c_2 = v_{c_2}^l \textbf{ AND...AND } c_{n_c} = v_{c_{nc}}^l \textbf{ THEN } d_1 = v_{d_1}^l$$
$$\textbf{AND } d_2 = v_{d_2}^l \textbf{ AND...AND } d_{n_d} = v_{d_{n_d}}^l .$$

Definition 8.3. The *P-indiscernibility relation refers* to a $\widetilde{P}$ relation defined in the space $U \times U$ satisfying

$$x_a \widetilde{P} x_b \iff \forall_q \in P;\ f_{x_a}(q) = f_{x_b}(q), \tag{8.1}$$

where $x_a, x_b \in U$, $P \subseteq Q$.

It is easy to verify that the relation $\widetilde{P}$ is reflexive, symmetrical and transitive, and thus it is a relation of equivalence. The relation of equivalence divides a set in which it is defined, into a family of disjoint sets called equivalence classes of this relation.

Definition 8.4. The set of all objects $x \in U$ being in relation $\widetilde{P}$ we call *the equivalence class* of relation $\widetilde{P}$ in the space U. For each $x_a \in U$, there is exactly one such set denoted by the symbol $[x_a]_{\widetilde{P}}$, i.e.

$$[x_a]_{\widetilde{P}} = \left\{ x \in U : x_a \widetilde{P} x \right\}. \tag{8.2}$$

The family of all equivalence classes of the relation $\widetilde{P}$ in the space U (called the quotient of set U by relation $\widetilde{P}$) will be denoted using the symbol P^* or $U/\widetilde{P}$. In the space U, certain sets X may exist. We can infer that particular objects $x \in U$ belong to sets X based on the knowledge of values of their features. The set of available features $P \subseteq Q$ is usually limited and the determination of membership of the object to a specific set may not be unequivocal. This situation is described by the terms of lower and upper approximation of set $X \subseteq U$.

Definition 8.5. The set $\underline{\widetilde{P}}X$ described as follows:

$$\underline{\widetilde{P}}X = \left\{ x \in U : [x]_{\widetilde{P}} \subseteq X \right\} \tag{8.3}$$

is called $\widetilde{P}$-lower approximation of the set $X \subseteq U$. Therefore, the lower approximation of the set X is the set of the objects $x \in U$, with relation to which on the basis of values of features P, we can certainly state that they are elements of the set X.

Definition 8.6. The set $\overline{\widetilde{P}}X$ described as follows:

$$\overline{\widetilde{P}}X = \left\{ x \in U : [x]_{\widetilde{P}} \cap x \neq \varnothing \right\} \tag{8.4}$$

is called *$\widetilde{P}$-upper approximation* of the set $X \subseteq U$. The upper approximation of the set X is the set of the objects $x \in U$, with relation to which, on the basis of values of features P, we can not certainly state that they are not elements of the set X.

Definition 8.7. The rough set X is defined as a pair of sets

$$(\underline{\widetilde{P}}X, \overline{\widetilde{P}}X) \tag{8.5}$$

The set X and its lower and upper approximation fulfill the inequality

$$\underline{\widetilde{P}}X \subseteq X \subseteq \overline{\widetilde{P}}X \tag{8.6}$$

8.2 Fuzzy Classifiers

All of systems constituting the soft computing concept have their advantages and disadvantages. Most applications utilize only one type of intelligent methods such as neural networks, fuzzy systems, rough sets or evolutionary computing. Neural networks have ability to perfectly fit to data. Fuzzy logic use interpretable knowledge and rough set systems can operate on data with missing feature values. By combining various soft computing methods we can achieve the synergy between them.

As we mentioned, one of soft computing applications is classification which consists in assigning an object described by a set of features to a class. The object $x \in \mathbf{X}$ is described by the vector of features $\mathbf{v} \in \mathbf{V}$. Thus we can equate object x class membership with its feature values $\overline{\mathbf{v}} = [\overline{v}_1, \overline{v}_2, \ldots, \overline{v}_n]$ class membership. Consequently, we can use interchangeably x or $\overline{\mathbf{v}}$. Let us assume that a fuzzy set $A \subseteq \mathbf{V}$ is given as its membership function $\mu_A(x) = \mu_A(\overline{\mathbf{v}}) = \mu_A(\overline{v}_1, \overline{v}_2, \ldots, \overline{v}_n)$ where $\overline{v}_i \in \mathbf{V}_i$ for $i = 1, \ldots, n$. We also define the set of all object x features $Q = \{v_1, v_2, \ldots, v_n\}$. There are many methods for classifying data. The traditional statistical classification procedures apply Bayesian decision theory and assume knowledge of the posterior probabilities. Unfortunately, in practical situations we have no information about an underlying probability model and Bayes formula cannot be applied. Over the years, numerous classification methods were developed [5] based on neural networks, fuzzy systems [7, 10, 14, 31], support vector machines, rough sets and other soft computing techniques. These methods do not need the information about the probability model. Yet, they are usually not applicable to classify correctly in the case of missing data (feature values).

Fuzzy classifiers are frequently used thanks to their ability to use knowledge in the form of intelligible IF–THEN fuzzy rules. The standard form of fuzzy rules are suitable for approximation and majority of control tasks. In the case of classification tasks, rules in other form are more appropriate, i.e.

$$\begin{array}{c} R^r: \text{IF } v_1 \text{ is } A_1^r \text{ AND } v_2 \text{ is } A_2^r \text{ AND} \ldots \\ \ldots \text{ AND } v_n \text{ is } A_n^r \\ \text{THEN } x \in \omega_1(\overline{z}_1^r), x \in \omega_2(\overline{z}_2^r), \ldots, x \in \omega_m(\overline{z}_m^r) \end{array}, \tag{8.7}$$

where $r = 1, \ldots, N$, N is the number of rules and $\overline{z}_j^r$ is the membership degree of the object x to the j–th class ω_j according to rule r. Let us assume that the membership of objects to classes is not fuzzy but crisp, i.e.

$$\overline{z}_j^r = \begin{cases} 1 & \text{if } x \in \omega_j \\ -1 & \text{if } x \notin \omega_j \end{cases}. \tag{8.8}$$

We write just $x \in \omega_j$ when $\overline{z}_j^r = 1$ (which means that object x belongs to the j–th class, according to the r–th rule) in definition of the r–th rule. We can omit the part $x \in \omega_j(\overline{z}_j^r)$ when $\overline{z}_j^r = -1$ (what means that object x does not belong to the j–th class, according to the r–th rule). In the case when values of all features v_i of classified object x are available, it is easy to adopt any formula of neuro-fuzzy system to classification tasks. They fall into one of the following categories [25], depending on the connective between the antecedent and the consequent in fuzzy rules: Takagi-Sugeno, Mamdani and logical type fuzzy systems. Logical type reasoning turned out to be better in classification tasks [24, 27, 28].

Classifiers can be combined to improve accuracy [11]. By combining intelligent learning systems, the model robustness and accuracy is nearly always improved, comparing to single-model solutions. Popular methods are bagging and boosting which are meta-algorithms for learning different classifiers. They assign weights to learning samples according to their performance on earlier classifiers in the ensemble. Thus subsystems are trained with different datasets created from the base dataset.

In this chapter we will combine fuzzy methods with the rough set theory [20, 21, 22] and classifier ensemble methods. An ensemble of neuro-fuzzy systems is trained with the AdaBoost algorithm and the backpropagation [9]. Then rules from neuro-fuzzy constituting the ensemble are used in a rough-neuro-fuzzy classifier. In this way we obtain rules that are perfectly fitted to data and use them in the classifier which can operate on data with missing feature values. Experiments on two well known data sets proved the effectiveness of the proposed solutions. The rough-neuro classifier performs very well, moreover the number of incorrect classifications does not increases with the number of missing feature values.

In this chapter we will study fuzzy classifiers based on the following fuzzy implications:

- S–implications
$$I(a,b) = S(N(a),b), \tag{8.9}$$
represented by Lukasiewicz, Reichenbach, Kleene-Dienes, Fodor and Dubois-Prade implications.
- R–implications
$$I(a,b) = \sup_{z\in[0,1]} \{z|T(a,z) \leq b\}, \tag{8.10}$$
represented by Rescher, Goguen and Gödel implications.
- QL–implications
$$I(a,b) = S(N(a),T(a,b)), \tag{8.11}$$
represented by Zadeh implication,
- D–implications [12]
$$I(a,b) = S(T(N(a),N(b)),b). \tag{8.12}$$

In the literature various neuro-fuzzy classifiers have been developed [15, 16, 17, 25, 26]. When we use the DCOG defuzzification and Mamdani approach to reasoning

$$\bar{z}_j = \frac{\sum_{r=1}^{N} \bar{z}_j^r \cdot \underset{k=1}{\overset{N}{S}} T\left(\mu_{A^k}(\mathbf{v}), \mu_{B_j^k}\left(\bar{z}_j^r\right)\right)}{\sum_{r=1}^{N} \underset{k=1}{\overset{N}{S}} T\left(\mu_{A^k}(\mathbf{v}), \mu_{B_j^k}\left(\bar{z}_j^r\right)\right)}. \tag{8.13}$$

When we use the DCOG defuzzification and logical approach to reasoning

$$\bar{z}_j = \frac{\sum_{r=1}^{N} \bar{z}_j^r \cdot \underset{k=1}{\overset{N}{T}} I\left(\mu_{A^k}(\mathbf{v}), \mu_{B_j^k}\left(\bar{z}_j^r\right)\right)}{\sum_{r=1}^{N} \underset{k=1}{\overset{N}{T}} I\left(\mu_{A^k}(\mathbf{v}), \mu_{B_j^k}\left(\bar{z}_j^r\right)\right)}. \tag{8.14}$$

When we use the MICOG defuzzification

$$\overline{z}_j = \frac{\sum_{r=1}^{N} \overline{z}_j^r \cdot g\left(\mu_{A^r}(\mathbf{v}), w_j^r\right)}{\sum_{r=1}^{N} g\left(\mu_{A^r}(\mathbf{v}), w_j^r\right)}. \tag{8.15}$$

where g is defined as follows

$$g\left(\mu_{A^r}(\bar{\mathbf{v}}), \bar{w}_j^r\right) = \int_{z_j \in \mathbf{Y}_j} \left(\mu_{B_j'^r}(z_j) - \alpha^r\right) dz_j \tag{8.16}$$

and $\bar{w}_j^r$ are parameters characterizing the width of fuzzy sets B_j^r, $j = 1,...,N$, $r = 1,...,N$ and parameter α^r determines the size of noninformative part of membership function of fuzzy sets B_j^r. Table 8.1 presents descriptions of various fuzzy classifiers depending on fuzzification, inference and defuzzification methods. We will assume that (8.8) holds and

(i)

$$\mu_{B_j^r}(z_j) = \begin{cases} 1 & \text{if } z_j = \overline{z}_j^r \\ 0 & \text{if } z_j = -\overline{z}_j^r, \end{cases} \tag{8.17}$$

or

(ii)

$$\mu_{B_j^r}(z_j) = \begin{cases} 1 & \text{if } z_j = \overline{z}_j^r \\ 0 & \text{if } z_j \neq \overline{z}_j^r, \end{cases} \tag{8.18}$$

or

(iii)

$$(iii)\mu_{B_j^r}(z_j) = z_j \stackrel{=}{*} \overline{z}_j^r \tag{8.19}$$

where

$$a \stackrel{=}{*} b = \begin{cases} 1 & \text{if } a = b \\ 0 & \text{if } b \neq b \end{cases} \tag{8.20}$$

Table 8.1 List of fuzzy classifiers

Fuzzification	Inference	Defuzzification	Abbreviation
singleton	t–norm	CA	T–CA
singleton	t–norm	DCOG	T–DCOG
singleton	Mamdani	MICOG	Mam–MICOG
singleton	Larsen	MICOG	Lar–MICOG
singleton	S/D-implication	DCOG	S/D–DCOG
singleton	R-implication	DCOG	R–DCOG
singleton	QL-implication	DCOG	QL–DCOG
singleton	Kleene–Dienes	MICOG	Kle-Die-MICOG
singleton	Łukasiewicz	MICOG	Łuk–MICOG
singleton	Reichenbach	MICOG	Rei–MICOG

Moreover, for future use we denote

$$a \overset{\neq}{*} b = \begin{cases} 1 & \text{if } a \neq b \\ 0 & \text{if } a = b \end{cases} \tag{8.21}$$

Using the above assumptions we can simplify particular groups of fuzzy implications.

- S–implications

$$I\left(\mu_{A^k}(\mathbf{v}), \mu_{B_j^k}\left(\overline{z}_j^r\right)\right) = S\left(N\left(\mu_{A^k}(\mathbf{v})\right), \mu_{B_j^k}\left(\overline{z}_j^r\right)\right), \tag{8.22}$$

$$I\left(\mu_{A^k}(\mathbf{v}), \mu_{B_j^k}\left(\overline{z}_j^r\right)\right) = S\left(N\left(\mu_{A^k}(\mathbf{v})\right), \overline{z}_j^k \overset{=}{*} \overline{z}_j^r\right), \tag{8.23}$$

$$I\left(\mu_{A^k}(\mathbf{v}), \mu_{B_j^k}\left(\overline{z}_j^r\right)\right) = N\left(T\left(\mu_{A^k}(\mathbf{v}), \overline{z}_j^k \overset{\neq}{*} \overline{z}_j^r\right)\right). \tag{8.24}$$

For the product t-norm, (8.24) becomes

$$I\left(\mu_{A^k}(\mathbf{v}), \mu_{B_j^k}\left(\overline{z}_j^r\right)\right) = N\left(\mu_{A^k}(\mathbf{v}) \cdot \left(\overline{z}_j^k \overset{\neq}{*} \overline{z}_j^r\right)\right), \tag{8.25}$$

or alternatively

$$I\left(\mu_{A^k}(\mathbf{v}), \mu_{B_j^k}\left(\overline{z}_j^r\right)\right) = \begin{cases} N\left(\mu_{A^k}(\mathbf{v})\right) & \text{if } \overline{z}_j^k \overset{=}{*} \overline{z}_j^r = 0 \\ 1 & \text{if } \overline{z}_j^k \overset{=}{*} \overline{z}_j^r = 1 \end{cases}, \tag{8.26}$$

$$I\left(\mu_{A^k}(\mathbf{v}), \mu_{B_j^k}\left(\overline{z}_j^r\right)\right) = N\left(\mu_{A^k}(\mathbf{v})\right) \cdot \left(\overline{z}_j^k \overset{\neq}{*} \overline{z}_j^r\right) + \left(\overline{z}_j^k \overset{=}{*} \overline{z}_j^r\right), \tag{8.27}$$

- R–implications

$$I\left(\mu_{A^k}(\mathbf{v}), \mu_{B_j^k}\left(\overline{z}_j^r\right)\right) = \sup_{z \in [0,1]} \left\{z | T\left(\mu_{A^k}(\mathbf{v}), z\right) \leq \mu_{B_j^k}\left(\overline{z}_j^r\right)\right\}, \tag{8.28}$$

$$I\left(\mu_{A^k}(\mathbf{v}), \mu_{B_j^k}\left(\overline{z}_j^r\right)\right) = \sup_{z \in [0,1]} \left\{z | T\left(\mu_{A^k}(\mathbf{v}), z\right) \leq \left(\overline{z}_j^k \overset{=}{*} \overline{z}_j^r\right)\right\}, \tag{8.29}$$

$$I\left(\mu_{A^k}(\mathbf{v}), \mu_{B_j^k}\left(\overline{z}_j^r\right)\right) = \begin{cases} 1 & \left(\overline{z}_j^k \overset{=}{*} \overline{z}_j^r\right) = 1 \\ 1 & \left(\overline{z}_j^k \overset{=}{*} \overline{z}_j^r\right) = 0 \text{ and } \mu_{\mathrm{A}^k}(\mathbf{v}) = 0 \\ 0 & \left(\overline{z}_j^k \overset{=}{*} \overline{z}_j^r\right) = 0 \text{ and } \mu_{\mathrm{A}^k}(\mathbf{v}) > 0 \end{cases}, \tag{8.30}$$

$$I\left(\mu_{A^k}(\mathbf{v}), \mu_{B_j^k}\left(\overline{z}_j^r\right)\right) = \left(\mu_{A^k}(\mathbf{v}) \overset{=}{*} 0\right) \cdot \left(\overline{z}_j^k \overset{\neq}{*} \overline{z}_j^r\right) + \left(\overline{z}_j^k \overset{=}{*} \overline{z}_j^r\right), \tag{8.31}$$

Let us note that expression $\mu_{A^k}(\mathbf{v}) \overset{=}{*} 0$ is a specific case of negation $N\left(\mu_{A^k}(\mathbf{v})\right)$. Therefore it is a special case of S-implication.

- QL–implications

$$I\left(\mu_{A^k}(\mathbf{v}),\mu_{B_j^k}\left(\overline{z}_j^r\right)\right)=S\left(N(\mu_{A^k}(\mathbf{v})),T(\mu_{A^k}(\mathbf{v}),\mu_{B_j^k}\left(\overline{z}_j^r\right))\right), \tag{8.32}$$

$$I\left(\mu_{A^k}(\mathbf{v}),\mu_{B_j^k}\left(\overline{z}_j^r\right)\right)=S\left(N(\mu_{A^k}(\mathbf{v})),T(\mu_{A^k}(\mathbf{v}),\overline{z}_j^k \overline{\overline{*}}\, \overline{z}_j^r)\right), \tag{8.33}$$

For the product t-norm, (8.33) becomes

$$I\left(\mu_{A^k}(\mathbf{v}),\mu_{B_j^k}\left(\overline{z}_j^r\right)\right)=S\left(N(\mu_{A^k}(\mathbf{v})),\mu_{A^k}(\mathbf{v})\cdot\left(\overline{z}_j^k \overline{\overline{*}}\, \overline{z}_j^r\right)\right), \tag{8.34}$$

- D–implications [12]

$$I\left(\mu_{A^k}(\mathbf{v}),\mu_{B_j^k}\left(\overline{z}_j^r\right)\right)=S\left(T\left(N(\mu_{A^k}(\mathbf{v})),N(\mu_{B_j^k}\left(\overline{z}_j^r\right))\right),\mu_{B_j^k}\left(\overline{z}_j^r\right)\right). \tag{8.35}$$

$$I\left(\mu_{A^k}(\mathbf{v}),\mu_{B_j^k}\left(\overline{z}_j^r\right)\right)=S\left(T\left(N(\mu_{A^k}(\mathbf{v})),N(\overline{z}_j^k \overline{\overline{*}}\, \overline{z}_j^r)\right),\overline{z}_j^k \overline{\overline{*}}\, \overline{z}_j^r\right). \tag{8.36}$$

It is easy to verify that

$$I\left(\mu_{A^k}(\mathbf{v}),\mu_{B_j^k}\left(\overline{z}_j^r\right)\right)=\begin{cases}N\left(\mu_{A^k}(\mathbf{v})\right) & \text{if } \overline{z}_j^k \overline{\overline{*}}\, \overline{z}_j^r=0\\ 1 & \text{if } \overline{z}_j^k \overline{\overline{*}}\, \overline{z}_j^r=1\end{cases}, \tag{8.37}$$

Therefore the systems with D–implications are identical to systems with S–implications. We will now present descriptions of systems listed in Table 8.1 using the notation introduced earlier in this chapter.

8.2.1 T–CA Fuzzy Classifier

The T–CA fuzzy classifier is given as follows

$$\overline{z}_j=\frac{\sum\limits_{r=1}^{N}\overline{z}_j^r\cdot\mu_{A^r}(\mathbf{v})}{\sum\limits_{r=1}^{N}\mu_{A^r}(\mathbf{v})}. \tag{8.38}$$

8.2.2 T–DCOG Fuzzy Classifier

Combining (8.13) with (8.20) we get the description of the T–DCOG fuzzy classifier

$$\overline{z}_j=\frac{\sum\limits_{r=1}^{N}\overline{z}_j^r\cdot\mathop{S}\limits_{k=1}^{N}\left[\left(\overline{z}_j^k \overline{\overline{*}}\, \overline{z}_j^r\right)\cdot\mu_{A^k}(\mathbf{v})\right]}{\sum\limits_{r=1}^{N}\mathop{S}\limits_{k=1}^{N}\left[\left(\overline{z}_j^k \overline{\overline{*}}\, \overline{z}_j^r\right)\cdot\mu_{A^k}(\mathbf{v})\right]}. \tag{8.39}$$

or

$$\overline{z}_j = \frac{2\sum\limits_{r=1}^{N} \overline{z'}^r_j \cdot \mathop{S}\limits_{k=1}^{N}\left[\overline{z'}^k_j\cdot\mu_{A^k}(\mathbf{v})\right]}{\sum\limits_{r=1}^{N} \overline{z'}^r_j\cdot \mathop{S}\limits_{k=1}^{N}\left[\overline{z'}^k_j\cdot\mu_{A^k}(\mathbf{v})\right] + \sum\limits_{r=1}^{N} \neg\overline{z'}^r_j\cdot\mathop{S}\limits_{k=1}^{N}\left[\neg\overline{z'}^k_j\cdot\mu_{A^k}(\mathbf{v})\right]} - 1. \tag{8.40}$$

where

$$z' = \frac{z+1}{2} \tag{8.41}$$

or

$$z' = \begin{cases} 0 & z=-1 \\ 1 & z=1 \end{cases} \tag{8.42}$$

8.2.3 S/D–DCOG Fuzzy Classifier

Combining (8.14) with (8.24) we get the description of the S/D–DCOG fuzzy classifier

$$\overline{z}_j = \frac{\sum\limits_{r=1}^{N} \overline{z}^r_j \cdot \mathop{T}\limits_{k=1}^{N}\left[N\left(T\left(\mu_{A^k}(\mathbf{v}), \overline{z}^k_j \overset{\neq}{*} \overline{z}^r_j\right)\right)\right]}{\sum\limits_{r=1}^{N} \mathop{T}\limits_{k=1}^{N}\left[N\left(T\left(\mu_{A^k}(\mathbf{v}), \overline{z}^k_j \overset{\neq}{*} \overline{z}^r_j\right)\right)\right]}. \tag{8.43}$$

or

$$\overline{z}_j = \frac{2\sum\limits_{r=1}^{N} \overline{z'}^r_j \cdot \mathop{T}\limits_{k=1}^{N} \neg\overline{z'}^k_j \cdot N\left(\mu_{A^k}(\mathbf{v})\right)}{\sum\limits_{r=1}^{N} \overline{z'}^r_j \cdot \mathop{T}\limits_{k=1}^{N} \neg\overline{z'}^k_j \cdot N\left(\mu_{A^k}(\mathbf{v})\right) + \sum\limits_{r=1}^{N} \neg\overline{z'}^r_j \cdot \mathop{T}\limits_{k=1}^{N} \overline{z'}^k_j \cdot N\left(\mu_{A^k}(\mathbf{v})\right)} - 1. \tag{8.44}$$

8.2.4 R–DCOG Fuzzy Classifier

Combining (8.14) with (8.31) we get the description of the R–DCOG fuzzy classifier

$$\overline{z}_j = \frac{\sum\limits_{r=1}^{N} \overline{z}^r_j \cdot \mathop{T}\limits_{k=1}^{N}\left[\left(\mu_{A^k}(\mathbf{v}) \overset{=}{*} 0\right)\cdot\left(\overline{z}^k_j \overset{\neq}{*} \overline{z}^r_j\right) + \left(\overline{z}^k_j \overset{=}{*} \overline{z}^r_j\right)\right]}{\sum\limits_{r=1}^{N} \mathop{T}\limits_{k=1}^{N}\left[\left(\mu_{A^k}(\mathbf{v}) \overset{=}{*} 0\right)\cdot\left(\overline{z}^k_j \overset{\neq}{*} \overline{z}^r_j\right) + \left(\overline{z}^k_j \overset{=}{*} \overline{z}^r_j\right)\right]}. \tag{8.45}$$

or

$$\overline{z}_j = \frac{2\sum\limits_{r=1}^{N} \overline{z'}^r_j \cdot \mathop{T}\limits_{k=1}^{N} \neg\overline{z'}^k_j \cdot \left(\mu_{A^k}(\mathbf{v}) \overset{=}{*} 0\right)}{\sum\limits_{r=1}^{N} \overline{z'}^r_j \cdot \mathop{T}\limits_{k=1}^{N} \neg\overline{z'}^k_j \cdot \left(\mu_{A^k}(\mathbf{v}) \overset{=}{*} 0\right) + \sum\limits_{r=1}^{N} \neg\overline{z'}^r_j \cdot \mathop{T}\limits_{k=1}^{N} \overline{z'}^k_j \cdot \left(\mu_{A^k}(\mathbf{v}) \overset{=}{*} 0\right)} - 1. \tag{8.46}$$

8.2.5 QL–DCOG Fuzzy Classifier

Combining (8.14) with (8.34) we get the description of the QL–DCOG fuzzy classifier

$$\bar{z}_j = \frac{\sum_{r=1}^{N} \bar{z}_j^r \cdot \mathop{T}_{k=1}^{N} S\left(N\left(\mu_{A^k}(\mathbf{v})\right), \left(\mu_{A^k}(\mathbf{v}) \cdot \left(\bar{z}_j^k \overset{=}{*} \bar{z}_j^r\right)\right)\right)}{\sum_{r=1}^{N} \mathop{T}_{k=1}^{N} S\left(N\left(\mu_{A^k}(\mathbf{v})\right), \left(\mu_{A^k}(\mathbf{v}) \cdot \left(\bar{z}_j^k \overset{=}{*} \bar{z}_j^r\right)\right)\right)}. \tag{8.47}$$

or

$$\bar{z}_j = \frac{2 \sum_{r=1}^{N} \overline{z'}_j^r \cdot T \begin{pmatrix} \mathop{T}_{k=1}^{N} \overline{z'}_j^k \cdot S\left(N\left(\mu_{A^k}(\mathbf{v})\right), \mu_{A^k}(\mathbf{v})\right), \\ \mathop{T}_{k=1}^{N} \neg \overline{z'}_j^k \cdot N\left(\mu_{A^k}(\mathbf{v})\right) \end{pmatrix}}{\begin{pmatrix} \sum_{r=1}^{N} \overline{z'}_j^r \cdot T \begin{pmatrix} \mathop{T}_{k=1}^{N} \overline{z'}_j^k \cdot S\left(N\left(\mu_{A^k}(\mathbf{v})\right), \mu_{A^k}(\mathbf{v})\right), \\ \mathop{T}_{k=1}^{N} \neg \overline{z'}_j^k \cdot N\left(\mu_{A^k}(\mathbf{v})\right) \end{pmatrix} \\ + \sum_{r=1}^{N} \neg \overline{z'}_j^r \cdot T \begin{pmatrix} \mathop{T}_{k=1}^{N} \neg \overline{z'}_j^k \cdot S\left(N\left(\mu_{A^k}(\mathbf{v})\right), \mu_{A^k}(\mathbf{v})\right), \\ \mathop{T}_{k=1}^{N} \overline{z'}_j^k \cdot N\left(\mu_{A^k}(\mathbf{v})\right) \end{pmatrix} \end{pmatrix}} - 1. \tag{8.48}$$

8.3 Rough Fuzzy Classifiers

In this section we will design various rough fuzzy classifiers assuming that some features are missing. As we indicated in Section 8.1, Q denotes a set of all features of object x. In the sequel, $P \subseteq Q$ denotes a set of features which values are known, and $G = Q \backslash P$ denotes the set of features which values are unknown. The construction of rough-fuzzy classifiers will be based on the following definition of the rough fuzzy set:

Definition 8.8 (Rough Fuzzy set). The rough fuzzy set is a pair $\{\underline{R}A, \overline{R}A\}$, where the set $\underline{R}A$ is R–lower aproximation of fuzzy set $A \subseteq U$, and the set $\overline{R}A$ is its R–upper approximation. Membership function of sets $\underline{R}A$ and $\overline{R}A$ are defined as follows:

$$\mu_{\underline{R}A}\left([\hat{x}]_R\right) = \inf_{x \in [\hat{x}]_R} \mu_A(x), \tag{8.49}$$

$$\mu_{\overline{R}A}\left([\hat{x}]_R\right) = \sup_{x \in [\hat{x}]_R} \mu_A(x). \tag{8.50}$$

The membership functions of lower and upper approximations of rough fuzzy set $\widetilde{P}A$ can be described by

$$\mu_{\underline{\widetilde{P}}A^r}(x) = \inf_{\mathbf{v}_G \in \mathbf{V}_G} \mu_A^r(\mathbf{v}_P, \mathbf{v}_G) \tag{8.51}$$

and

$$\mu_{\overline{\widetilde{P}}A^r}(x) = \sup_{\mathbf{v}_G \in \mathbf{V}_G} \mu_A^r(\mathbf{v}_P, \mathbf{v}_G). \tag{8.52}$$

If we assume that fuzzy set A^r, $r = 1,...,N$ is a Cartesian product, i.e. $A^r = A_1^r \times A_2^r \times ... \times A_n^r$, then the membership function of its $\widetilde{P}$–lower approximation is given by the following equation

$$\mu_{\underline{\widetilde{P}}A^r}(x) = T\left(\underset{i:v_i \in P}{T} \mu_{A_i^r}(v_i), \underset{i:v_i \in G}{T} \inf_{v_i \in V_i} \mu_{A_i^r}(v_i) \right), \tag{8.53}$$

The membership function of its $\widetilde{P}$–upper approximation is given by the following equation

$$\mu_{\overline{\widetilde{P}}A^r}(x) = T\left(\underset{i:v_i \in P}{T} \mu_{A_i^r}(v_i), \underset{i:v_i \in G}{T} \sup_{v_i \in V_i} \mu_{A_i^r}(v_i) \right). \tag{8.54}$$

Formulas (8.53) and (8.54) take into consideration vector of only known values of input features i.e. $v_i \in P$ and the infinium and supremum of memberships of unknown features, i.e. $v_i \in G$. In order to prove the theorems presented in this section we will use arguments analogical to those presented by Nowicki in [16].

8.3.1 T–CA Rough Fuzzy Classifier

The lower and upper approximations of the membership of object x to class ω_j are given by

$$\underline{z}_j = \frac{\sum_{r=1}^{N} \overline{z}_j^r \cdot \mu_{A_L^r}(\mathbf{v})}{\sum_{r=1}^{N} \mu_{A_L^r}(\mathbf{v})}, \tag{8.55}$$

$$\overline{z}_j = \frac{\sum_{r=1}^{N} \overline{z}_j^r \cdot \mu_{A_U^r}(\mathbf{v})}{\sum_{r=1}^{N} \mu_{A_U^r}(\mathbf{v})}. \tag{8.56}$$

Theorem 8.1. *Fuzzy sets A_L^r and A_U^r in descriptions* (8.55) *and* (8.56) *should be chosen as follows*

$$A_L^r = \begin{cases} \underline{\widetilde{P}}A^r & \text{if } \overline{z}_j^r = 1 \\ \overline{\widetilde{P}}A^r & \text{if } \overline{z}_j^r = -1. \end{cases} \tag{8.57}$$

$$A_U^r = \begin{cases} \overline{\widetilde{P}}A^r & \text{if } \overline{z}_j^r = 1 \\ \underline{\widetilde{P}}A^r & \text{if } \overline{z}_j^r = -1. \end{cases} \tag{8.58}$$

Proof. We will find sign of derivatives of expressions (8.55) and (8.56) with respect to $\mu_{A^l_\mathrm{L}}$ and $\mu_{A^l_\mathrm{U}}$, for l such that $\overline{z}^l_j = -1$ and for l such that $\overline{z}^l_j = 1$. A simple algebra leads to the following formulas

$$\left.\frac{\partial \underline{\overline{z}}_j}{\partial \mu_{A^l_\mathrm{L}}(\mathbf{v})}\right|_{l:\, \overline{z}^l_j=-1} = \frac{\partial}{\partial \mu_{A^l_\mathrm{L}}(\mathbf{v})} \left.\frac{\sum\limits_{r=1}^{N} \overline{z}^r_j \cdot \mu_{A^r_\mathrm{L}}(\mathbf{v})}{\sum\limits_{r=1}^{N} \mu_{A^r_\mathrm{L}}(\mathbf{v})}\right|_{l:\, \overline{z}^l_j=-1}, \tag{8.59}$$

$$\left.\frac{\partial \underline{\overline{z}}_j}{\partial \mu_{A^l_\mathrm{L}}(\mathbf{v})}\right|_{l:\, \overline{z}^l_j=-1} = \frac{\left.\frac{\partial \sum\limits_{r=1}^{N} \overline{z}^r_j \cdot \mu_{A^r_\mathrm{L}}(\mathbf{v})}{\partial \mu_{A^l_\mathrm{L}}(\mathbf{v})}\right|_{l:\, \overline{z}^l_j=-1} \cdot \sum\limits_{r=1}^{N} \mu_{A^r_\mathrm{L}}(\mathbf{v}) - \sum\limits_{r=1}^{N} \overline{z}^r_j \cdot \mu_{A^r_\mathrm{L}}(\mathbf{v}) \cdot \left.\frac{\partial \sum\limits_{r=1}^{N} \mu_{A^r_\mathrm{L}}(\mathbf{v})}{\partial \mu_{A^l_\mathrm{L}}(\mathbf{v})}\right|_{l:\, \overline{z}^l_j=-1}}{\left[\sum\limits_{r=1}^{N} \mu_{A^r_\mathrm{L}}(\mathbf{v})\right]^2}, \tag{8.60}$$

$$\left.\frac{\partial \underline{\overline{z}}_j}{\partial \mu_{A^l_\mathrm{L}}(\mathbf{v})}\right|_{l:\, \overline{z}^l_j=-1} = \frac{-1 \sum\limits_{r=1}^{N} \mu_{A^r_\mathrm{L}}(\mathbf{v}) - \sum\limits_{r=1}^{N} \overline{z}^r_j \cdot \mu_{A^r_\mathrm{L}}(\mathbf{v}) \cdot 1}{\left[\sum\limits_{r=1}^{N} \mu_{A^r_\mathrm{L}}(\mathbf{v})\right]^2}, \tag{8.61}$$

$$\left.\frac{\partial \underline{\overline{z}}_j}{\partial \mu_{A^l_\mathrm{L}}(\mathbf{v})}\right|_{l:\, \overline{z}^l_j=-1} = \frac{-\sum\limits_{r=1}^{N} \left(1 + \overline{z}^r_j\right) \cdot \mu_{A^r_\mathrm{L}}(\mathbf{v})}{\left[\sum\limits_{r=1}^{N} \mu_{A^r_\mathrm{L}}(\mathbf{v})\right]^2}, \tag{8.62}$$

Since $\mu_{A^r}(\mathbf{v}) \geq 0$ and $\overline{z}^r_j \in \{-1, 1\}$, we get

$$\left.\frac{\partial \underline{\overline{z}}_j}{\partial \mu_{A^l_\mathrm{L}}(\mathbf{v})}\right|_{l:\, \overline{z}^l_j=-1} \leq 0. \tag{8.63}$$

By analogy we obtain

$$\left.\frac{\partial \overline{\overline{z}}_j}{\partial \mu_{A^l_\mathrm{U}}(\mathbf{v})}\right|_{l:\, \overline{z}^l_j=-1} = \frac{\partial}{\partial \mu_{A^l_\mathrm{U}}(\mathbf{v})} \left.\frac{\sum\limits_{r=1}^{N} \overline{z}^r_j \cdot \mu_{A^r_\mathrm{U}}(\mathbf{v})}{\sum\limits_{r=1}^{N} \mu_{A^r_\mathrm{U}}(\mathbf{v})}\right|_{l:\, \overline{z}^l_j=-1}, \tag{8.64}$$

$$\left.\frac{\partial \overline{\overline{z}}_j}{\partial \mu_{A^l_\mathrm{U}}(\mathbf{v})}\right|_{l:\, \overline{z}^l_j=-1} = \frac{\left.\frac{\partial \sum\limits_{r=1}^{N} \overline{z}^r_j \cdot \mu_{A^r_\mathrm{L}}(\mathbf{v})}{\partial \mu_{A^l_\mathrm{U}}(\mathbf{v})}\right|_{l:\, \overline{z}^l_j=-1} \cdot \sum\limits_{r=1}^{N} \mu_{A^r_\mathrm{U}}(\mathbf{v}) - \sum\limits_{r=1}^{N} \overline{z}^r_j \cdot \mu_{A^r_\mathrm{U}}(\mathbf{v}) \cdot \left.\frac{\partial \sum\limits_{r=1}^{N} \mu_{A^r_\mathrm{L}}(\mathbf{v})}{\partial \mu_{A^l_\mathrm{U}}(\mathbf{v})}\right|_{l:\, \overline{z}^l_j=-1}}{\left[\sum\limits_{r=1}^{N} \mu_{A^r_\mathrm{U}}(\mathbf{v})\right]^2}, \tag{8.65}$$

$$\left.\frac{\partial \overline{z}_j}{\partial \mu_{A^l_{\mathrm{U}}}(\mathbf{v})}\right|_{l:\, \overline{z}^l_j=-1} = \frac{-1 \sum\limits_{r=1}^{N} \mu_{A^r_{\mathrm{U}}}(\mathbf{v}) - \sum\limits_{r=1}^{N} \overline{z}^r_j \cdot \mu_{A^r_{\mathrm{U}}}(\mathbf{v}) \cdot 1}{\left[\sum\limits_{r=1}^{N} \mu_{A^r_{\mathrm{U}}}(\mathbf{v})\right]^2}, \tag{8.66}$$

$$\left.\frac{\partial \overline{z}_j}{\partial \mu_{A^l_{\mathrm{U}}}(\mathbf{v})}\right|_{l:\, \overline{z}^l_j=-1} = \frac{-\sum\limits_{r=1}^{N} \left(1 + \overline{z}^r_j\right) \cdot \mu_{A^r_{\mathrm{U}}}(\mathbf{v})}{\left[\sum\limits_{r=1}^{N} \mu_{A^r_{\mathrm{U}}}(\mathbf{v})\right]^2}, \tag{8.67}$$

Since $\mu_{A^r}(\mathbf{v}) \geq 0$ and $\overline{z}^r_j \in \{-1, 1\}$, we obtain

$$\left.\frac{\partial \overline{z}_j}{\partial \mu_{A^l_{\mathrm{U}}}(\mathbf{v})}\right|_{l:\, \overline{z}^l_j=-1} \leq 0. \tag{8.68}$$

For $l\colon \overline{z}^l_j = 1$, that is for a fuzzy rule with $x \in \omega_j$ in its consequent, we have a sequence of analogical expressions

$$\left.\frac{\partial \underline{z}_j}{\partial \mu_{A^l_{\mathrm{L}}}(\mathbf{v})}\right|_{l:\, \overline{z}^l_j=1} = \left.\frac{\partial}{\partial \mu_{A^l_{\mathrm{L}}}(\mathbf{v})} \frac{\sum\limits_{r=1}^{N} \overline{z}^r_j \cdot \mu_{A^r_{\mathrm{L}}}(\mathbf{v})}{\sum\limits_{r=1}^{N} \mu_{A^r_{\mathrm{L}}}(\mathbf{v})}\right|_{l:\, \overline{z}^l_j=1}, \tag{8.69}$$

$$\left.\frac{\partial \underline{z}_j}{\partial \mu_{A^l_{\mathrm{L}}}(\mathbf{v})}\right|_{l:\, \overline{z}^l_j=1} = \frac{\left.\frac{\partial \sum\limits_{r=1}^{N} \overline{z}^r_j \cdot \mu_{A^r_{\mathrm{L}}}(\mathbf{v})}{\partial \mu_{A^l_{\mathrm{L}}}(\mathbf{v})}\right|_{l:\, \overline{z}^l_j=1} \cdot \sum\limits_{r=1}^{N} \mu_{A^r_{\mathrm{L}}}(\mathbf{v}) - \sum\limits_{r=1}^{N} \overline{z}^r_j \cdot \mu_{A^r_{\mathrm{L}}}(\mathbf{v}) \cdot \left.\frac{\partial \sum\limits_{r=1}^{N} \mu_{A^r_{\mathrm{L}}}(\mathbf{v})}{\partial \mu_{A^l_{\mathrm{L}}}(\mathbf{v})}\right|_{l:\, \overline{z}^l_j=1}}{\left[\sum\limits_{r=1}^{N} \mu_{A^r_{\mathrm{L}}}(\mathbf{v})\right]^2}, \tag{8.70}$$

$$\left.\frac{\partial \underline{z}_j}{\partial \mu_{A^l_{\mathrm{L}}}(\mathbf{v})}\right|_{l:\, \overline{z}^l_j=1} = \frac{1 \sum\limits_{r=1}^{N} \mu_{A^r_{\mathrm{L}}}(\mathbf{v}) - \sum\limits_{r=1}^{N} \overline{z}^r_j \cdot \mu_{A^r_{\mathrm{L}}}(\mathbf{v}) \cdot 1}{\left[\sum\limits_{r=1}^{N} \mu_{A^r_{\mathrm{L}}}(\mathbf{v})\right]^2}, \tag{8.71}$$

$$\left.\frac{\partial \underline{z}_j}{\partial \mu_{A^l_{\mathrm{L}}}(\mathbf{v})}\right|_{l:\, \overline{z}^l_j=1} = \frac{\sum\limits_{r=1}^{N} \left(1 - \overline{z}^r_j\right) \cdot \mu_{A^r_{\mathrm{L}}}(\mathbf{v})}{\left[\sum\limits_{r=1}^{N} \mu_{A^r_{\mathrm{L}}}(\mathbf{v})\right]^2}, \tag{8.72}$$

Since $\mu_{A^r}(\mathbf{v}) \geq 0$ and $\overline{z}_j^r \in \{-1,1\}$, we can state that

$$\left.\frac{\partial \underline{z}_j}{\partial \mu_{A_{\mathrm{L}}^l}(\mathbf{v})}\right|_{l:\,\overline{z}_j^l=1} \geq 0. \tag{8.73}$$

Next we have

$$\left.\frac{\partial \overline{z}_j}{\partial \mu_{A_{\mathrm{U}}^l}(\mathbf{v})}\right|_{l:\,\overline{z}_j^l=1} = \left.\frac{\partial}{\partial \mu_{A_{\mathrm{U}}^l}(\mathbf{v})} \frac{\sum\limits_{r=1}^{N} \overline{z}_j^r \cdot \mu_{A_{\mathrm{U}}^r}(\mathbf{v})}{\sum\limits_{r=1}^{N} \mu_{A_{\mathrm{U}}^r}(\mathbf{v})}\right|_{l:\,\overline{z}_j^l=1}, \tag{8.74}$$

$$\left.\frac{\partial \overline{z}_j}{\partial \mu_{A_{\mathrm{U}}^l}(\mathbf{v})}\right|_{l:\,\overline{z}_j^l=1} = \frac{\left.\frac{\partial \sum\limits_{r=1}^{N} \overline{z}_j^r \cdot \mu_{A_{\mathrm{L}}^r}(\mathbf{v})}{\partial \mu_{A_{\mathrm{U}}^l}(\mathbf{v})}\right|_{l:\,\overline{z}_j^l=1} \cdot \sum\limits_{r=1}^{N} \mu_{A_{\mathrm{U}}^r}(\mathbf{v}) - \sum\limits_{r=1}^{N} \overline{z}_j^r \cdot \mu_{A_{\mathrm{U}}^r}(\mathbf{v}) \cdot \left.\frac{\partial \sum\limits_{r=1}^{N} \mu_{A_{\mathrm{L}}^r}(\mathbf{v})}{\partial \mu_{A_{\mathrm{U}}^l}(\mathbf{v})}\right|_{l:\,\overline{z}_j^l=1}}{\left[\sum\limits_{r=1}^{N} \mu_{A_{\mathrm{U}}^r}(\mathbf{v})\right]^2}, \tag{8.75}$$

$$\left.\frac{\partial \overline{z}_j}{\partial \mu_{A_{\mathrm{U}}^l}(\mathbf{v})}\right|_{l:\,\overline{z}_j^l=1} = \frac{1 \sum\limits_{r=1}^{N} \mu_{A_{\mathrm{U}}^r}(\mathbf{v}) - \sum\limits_{r=1}^{N} \overline{z}_j^r \cdot \mu_{A_{\mathrm{U}}^r}(\mathbf{v}) \cdot 1}{\left[\sum\limits_{r=1}^{N} \mu_{A_{\mathrm{U}}^r}(\mathbf{v})\right]^2}, \tag{8.76}$$

$$\left.\frac{\partial \overline{z}_j}{\partial \mu_{A_{\mathrm{U}}^l}(\mathbf{v})}\right|_{l:\,\overline{z}_j^l=1} = \frac{\sum\limits_{r=1}^{N} \left(1 - \overline{z}_j^r\right) \cdot \mu_{A_{\mathrm{U}}^r}(\mathbf{v})}{\left[\sum\limits_{r=1}^{N} \mu_{A_{\mathrm{U}}^r}(\mathbf{v})\right]^2}, \tag{8.77}$$

Since $\mu_{A^r}(\mathbf{v}) \geq 0$ and $\overline{z}_j^r \in \{-1,1\}$, we can conclude that

$$\left.\frac{\partial \overline{z}_j}{\partial \mu_{A_{\mathrm{U}}^l}(\mathbf{v})}\right|_{l:\,\overline{z}_j^l=1} \geq 0. \tag{8.78}$$

Analyzing inequalities (8.63), (8.68), (8.73) and (8.78) it is easily seen that in formulas (8.55) and (8.56) fuzzy sets A_{L}^r and A_{U}^r should be chosen according to descriptions (8.57) and (8.58).

8.3.2 T-DCOG Rough Fuzzy Classifier

Following formula (8.39), the lower and upper approximations of the membership of object x to class ω_j are given by

$$\underline{z}_j = \frac{\sum\limits_{r=1}^{N} \overline{z}_j^r \cdot \mathop{S}\limits_{k=1}^{N} \left[\left(\overline{z}_j^k \overset{=}{\ast} \overline{z}_j^r\right) \cdot \mu_{A_{\mathrm{L}}^k}(\mathbf{v})\right]}{\sum\limits_{r=1}^{N} \mathop{S}\limits_{k=1}^{N} \left[\left(\overline{z}_j^k \overset{=}{\ast} \overline{z}_j^r\right) \cdot \mu_{A_{\mathrm{L}}^k}(\mathbf{v})\right]}, \tag{8.79}$$

$$\overline{z}_j = \frac{\sum\limits_{r=1}^{N} \overline{z}_j^r \cdot \mathop{S}\limits_{k=1}^{N} \left[\left(\overline{z}_j^k \overset{=}{\ast} \overline{z}_j^r\right) \cdot \mu_{A_{\mathrm{U}}^k}(\mathbf{v})\right]}{\sum\limits_{r=1}^{N} \mathop{S}\limits_{k=1}^{N} \left[\left(\overline{z}_j^k \overset{=}{\ast} \overline{z}_j^r\right) \cdot \mu_{A_{\mathrm{U}}^k}(\mathbf{v})\right]}. \tag{8.80}$$

Theorem 8.2. *Fuzzy sets A_{L}^r and A_{U}^r in descriptions* (8.79) *and* (8.80) *should be chosen as follows*

$$A_{\mathrm{L}}^r = \begin{cases} \underline{\widetilde{P}}A^r & \text{if } \overline{\mathrm{z}}_{\mathrm{j}}^{\mathrm{r}} = 1 \\ \overline{\widetilde{P}}A^r & \text{if } \overline{\mathrm{z}}_{\mathrm{j}}^{\mathrm{r}} = -1. \end{cases} \tag{8.81}$$

$$A_{\mathrm{U}}^r = \begin{cases} \overline{\widetilde{P}}A^r & \text{if } \overline{\mathrm{z}}_{\mathrm{j}}^{\mathrm{r}} = 1 \\ \underline{\widetilde{P}}A^r & \text{if } \overline{\mathrm{z}}_{\mathrm{j}}^{\mathrm{r}} = -1. \end{cases} \tag{8.82}$$

Proof. In the first step for (8.79) and l such that $\overline{z}_j^l = -1$ we have

$$\left.\frac{\partial \underline{z}_j}{\partial \mu_{A_{\mathrm{L}}^l}(\mathbf{v})}\right|_{l:\, \overline{z}_j^l=-1} = \left.\frac{\partial}{\partial \mu_{A_{\mathrm{L}}^l}(\mathbf{v})} \frac{\sum\limits_{r=1}^{N} \overline{z}_j^r \cdot \mathop{S}\limits_{k=1}^{N} \left[\left(\overline{z}_j^k \overset{=}{\ast} \overline{z}_j^r\right) \cdot \mu_{A_{\mathrm{L}}^k}(\mathbf{v})\right]}{\sum\limits_{r=1}^{N} \mathop{S}\limits_{k=1}^{N} \left[\left(\overline{z}_j^k \overset{=}{\ast} \overline{z}_j^r\right) \cdot \mu_{A_{\mathrm{L}}^k}(\mathbf{v})\right]}\right|_{l:\, \overline{z}_j^l=-1}, \tag{8.83}$$

$$\left.\frac{\partial \underline{z}_j}{\partial \mu_{A_{\mathrm{L}}^l}(\mathbf{v})}\right|_{l:\, \overline{z}_j^l=-1} = \frac{\left(\begin{array}{l} \left.\frac{\partial \sum\limits_{r=1}^{N} \overline{z}_j^r \cdot \mathop{S}\limits_{k=1}^{N} \left[\left(\overline{z}_j^k \overset{=}{\ast} \overline{z}_j^r\right) \cdot \mu_{A_{\mathrm{L}}^k}(\mathbf{v})\right]}{\partial \mu_{A_{\mathrm{L}}^l}(\mathbf{v})}\right|_{l:\, \overline{z}_j^l=-1} \cdot \sum\limits_{r=1}^{N} \mathop{S}\limits_{k=1}^{N} \left[\left(\overline{z}_j^k \overset{=}{\ast} \overline{z}_j^r\right) \cdot \mu_{A_{\mathrm{L}}^k}(\mathbf{v})\right] - \\ \sum\limits_{r=1}^{N} \overline{z}_j^r \cdot \mathop{S}\limits_{k=1}^{N} \left[\left(\overline{z}_j^k \overset{=}{\ast} \overline{z}_j^r\right) \cdot \mu_{A_{\mathrm{L}}^k}(\mathbf{v})\right] \cdot \left.\frac{\partial \sum\limits_{r=1}^{N} \mathop{S}\limits_{k=1}^{N} \left[\left(\overline{z}_j^k \overset{=}{\ast} \overline{z}_j^r\right) \cdot \mu_{A_{\mathrm{L}}^k}(\mathbf{v})\right]}{\partial \mu_{A_{\mathrm{L}}^l}(\mathbf{v})}\right|_{l:\, \overline{z}_j^l=-1} \end{array}\right)}{\left[\sum\limits_{r=1}^{N} \mathop{S}\limits_{k=1}^{N} \left[\left(\overline{z}_j^k \overset{=}{\ast} \overline{z}_j^r\right) \cdot \mu_{A_{\mathrm{L}}^k}(\mathbf{v})\right]\right]^2}, \tag{8.84}$$

$$\left.\frac{\partial \underline{z}_j}{\partial \mu_{A_{\mathrm{L}}^l}(\mathbf{v})}\right|_{l:\, \overline{z}_j^l=-1} = \frac{\left(\begin{array}{l} \sum\limits_{r=1}^{N} \overline{z}_j^r \cdot \left.\frac{\partial \mathop{S}\limits_{k=1}^{N} \left[\left(\overline{z}_j^k \overset{=}{\ast} \overline{z}_j^r\right) \cdot \mu_{A_{\mathrm{L}}^k}(\mathbf{v})\right]}{\partial \mu_{A_{\mathrm{L}}^l}(\mathbf{v})}\right|_{l:\, \overline{z}_j^l=-1} \cdot \sum\limits_{r=1}^{N} \mathop{S}\limits_{k=1}^{N} \left[\left(\overline{z}_j^k \overset{=}{\ast} \overline{z}_j^r\right) \cdot \mu_{A_{\mathrm{L}}^k}(\mathbf{v})\right] - \\ \sum\limits_{r=1}^{N} \overline{z}_j^r \cdot \mathop{S}\limits_{k=1}^{N} \left[\left(\overline{z}_j^k \overset{=}{\ast} \overline{z}_j^r\right) \cdot \mu_{A_{\mathrm{L}}^k}(\mathbf{v})\right] \cdot \sum\limits_{r=1}^{N} \left.\frac{\partial \mathop{S}\limits_{k=1}^{N} \left[\left(\overline{z}_j^k \overset{=}{\ast} \overline{z}_j^r\right) \cdot \mu_{A_{\mathrm{L}}^k}(\mathbf{v})\right]}{\partial \mu_{A_{\mathrm{L}}^l}(\mathbf{v})}\right|_{l:\, \overline{z}_j^l=-1} \end{array}\right)}{\left[\sum\limits_{r=1}^{N} \mathop{S}\limits_{k=1}^{N} \left[\left(\overline{z}_j^k \overset{=}{\ast} \overline{z}_j^r\right) \cdot \mu_{A_{\mathrm{L}}^k}(\mathbf{v})\right]\right]^2}, \tag{8.85}$$

On the basis of boundary conditions of t-conorms, we get

$$\left.\frac{\partial \overline{z}_j}{\partial \mu_{A_L^l}(\mathbf{v})}\right|_{l:\,\overline{z}_j^l=-1} = \frac{\left(\begin{array}{c} \left(\sum\limits_{\substack{r=1\\ r:\,\overline{z}_j^r=1}}^{N} \left.\frac{\partial \underset{\substack{k=1\\ k:\,\overline{z}_j^k=1}}{\overset{N}{S}} \left[\mu_{A_L^k}(\mathbf{v})\right]}{\partial \mu_{A_L^l}(\mathbf{v})}\right|_{l:\,\overline{z}_j^l=-1} - \sum\limits_{\substack{r=1\\ r:\,\overline{z}_j^r=-1}}^{N} \left.\frac{\partial \underset{\substack{k=1\\ :\,\overline{z}_j^k=-1}}{\overset{N}{S}} \left[\mu_{A_L^k}(\mathbf{v})\right]}{\partial \mu_{A_L^l}(\mathbf{v})}\right|_{l:\,\overline{z}_j^l=-1} \right) \cdot \\ \cdot \sum\limits_{r=1}^{N} \underset{k=1}{\overset{N}{S}} \left[\left(\overline{z}_j^k \stackrel{=}{*} \overline{z}_j^r\right) \cdot \mu_{A_L^k}(\mathbf{v})\right] - \sum\limits_{r=1}^{N} \overline{z}_j^r \cdot \underset{k=1}{\overset{N}{S}} \left[\left(\overline{z}_j^k \stackrel{=}{*} \overline{z}_j^r\right) \cdot \mu_{A_L^k}(\mathbf{v})\right] \cdot \\ \cdot \left(\sum\limits_{\substack{r=1\\ \overline{z}_j^r=1}}^{N} \left.\frac{\partial \underset{\substack{k=1\\ \overline{z}_j^k=1}}{\overset{N}{S}} \left[\mu_{A_L^k}(\mathbf{v})\right]}{\partial \mu_{A_L^l}(\mathbf{v})}\right|_{l:\,\overline{z}_j^l=-1} + \sum\limits_{\substack{r=1\\ \overline{z}_j^r=-1}}^{N} \left.\frac{\partial \underset{\substack{k=1\\ \overline{z}_j^k=-1}}{\overset{N}{S}} \left[\mu_{A_L^k}(\mathbf{v})\right]}{\partial \mu_{A_L^l}(\mathbf{v})}\right|_{l:\,\overline{z}_j^l=-1} \right) \end{array}\right)}{\left[\sum\limits_{r=1}^{N} \underset{k=1}{\overset{N}{S}} \left[\left(\overline{z}_j^k \stackrel{=}{*} \overline{z}_j^r\right) \cdot \mu_{A_L^k}(\mathbf{v})\right]\right]^2}, \tag{8.86}$$

$$\left.\frac{\partial \overline{z}_j}{\partial \mu_{A_L^l}(\mathbf{v})}\right|_{l:\,\overline{z}_j^l=-1} = \frac{\left(\begin{array}{c} \left(0 - \sum\limits_{\substack{r=1\\ r:\,\overline{z}_j^r=-1}}^{N} \left.\frac{\partial \underset{\substack{k=1\\ :\,\overline{z}_j^k=-1}}{\overset{N}{S}} \left[\mu_{A_L^k}(\mathbf{v})\right]}{\partial \mu_{A_L^l}(\mathbf{v})}\right|_{l:\,\overline{z}_j^l=-1} \right) \cdot \\ \cdot \sum\limits_{r=1}^{N} \underset{k=1}{\overset{N}{S}} \left[\left(\overline{z}_j^k \stackrel{=}{*} \overline{z}_j^r\right) \cdot \mu_{A_L^k}(\mathbf{v})\right] - \sum\limits_{r=1}^{N} \overline{z}_j^r \cdot \underset{k=1}{\overset{N}{S}} \left[\left(\overline{z}_j^k \stackrel{=}{*} \overline{z}_j^r\right) \cdot \mu_{A_L^k}(\mathbf{v})\right] \cdot \\ \cdot \left(0 + \sum\limits_{\substack{r=1\\ \overline{z}_j^r=-1}}^{N} \left.\frac{\partial \underset{\substack{k=1\\ \overline{z}_j^k=-1}}{\overset{N}{S}} \left[\mu_{A_L^k}(\mathbf{v})\right]}{\partial \mu_{A_L^l}(\mathbf{v})}\right|_{l:\,\overline{z}_j^l=-1} \right) \end{array}\right)}{\left[\sum\limits_{r=1}^{N} \underset{k=1}{\overset{N}{S}} \left[\left(\overline{z}_j^k \stackrel{=}{*} \overline{z}_j^r\right) \cdot \mu_{A_L^k}(\mathbf{v})\right]\right]^2}, \tag{8.87}$$

$$\left.\frac{\partial \overline{z}_j}{\partial \mu_{A_L^l}(\mathbf{v})}\right|_{l:\,\overline{z}_j^l=-1} = \frac{\left(\begin{array}{c} - \sum\limits_{\substack{r=1\\ r:\,\overline{z}_j^r=-1}}^{N} \left.\frac{\partial \underset{\substack{k=1\\ :\,\overline{z}_j^k=-1}}{\overset{N}{S}} \left[\mu_{A_L^k}(\mathbf{v})\right]}{\partial \mu_{A_L^l}(\mathbf{v})}\right|_{l:\,\overline{z}_j^l=-1} \cdot \sum\limits_{r=1}^{N} \underset{k=1}{\overset{N}{S}} \left[\left(\overline{z}_j^k \stackrel{=}{*} \overline{z}_j^r\right) \cdot \mu_{A_L^k}(\mathbf{v})\right] \\ - \sum\limits_{r=1}^{N} \overline{z}_j^r \cdot \underset{k=1}{\overset{N}{S}} \left[\left(\overline{z}_j^k \stackrel{=}{*} \overline{z}_j^r\right) \cdot \mu_{A_L^k}(\mathbf{v})\right] \cdot \sum\limits_{\substack{r=1\\ \overline{z}_j^r=-1}}^{N} \left.\frac{\partial \underset{\substack{k=1\\ \overline{z}_j^k=-1}}{\overset{N}{S}} \left[\mu_{A_L^k}(\mathbf{v})\right]}{\partial \mu_{A_L^l}(\mathbf{v})}\right|_{l:\,\overline{z}_j^l=-1} \end{array}\right)}{\left[\sum\limits_{r=1}^{N} \underset{k=1}{\overset{N}{S}} \left[\left(\overline{z}_j^k \stackrel{=}{*} \overline{z}_j^r\right) \cdot \mu_{A_L^k}(\mathbf{v})\right]\right]^2}, \tag{8.88}$$

$$\left.\frac{\partial \underline{\overline{z}}_j}{\partial \mu_{A_{\mathrm{L}}^l}(\mathbf{v})}\right|_{l:\,\overline{z}_j^l=-1} = -\frac{\left(\sum_{\substack{r=1\\ r:\,\overline{z}_j^r=-1}}^{N} \left.\frac{\partial \underset{\substack{k=1\\ :\,\overline{z}_j^k=-1}}{\overset{N}{S}} \left[\mu_{A_{\mathrm{L}}^k}(\mathbf{v})\right]}{\partial \mu_{A_{\mathrm{L}}^l}(\mathbf{v})}\right|_{l:\,\overline{z}_j^l=-1} \cdot \sum_{r=1}^{N}\left(\overline{z}_j^r+1\right) \overset{N}{\underset{k=1}{S}}\left[\left(\overline{z}_j^k \,\overline{*}\, \overline{z}_j^r\right)\cdot \mu_{A_{\mathrm{L}}^k}(\mathbf{v})\right]\right)}{\left[\sum_{r=1}^{N} \overset{N}{\underset{k=1}{S}}\left[\left(\overline{z}_j^k \,\overline{*}\, \overline{z}_j^r\right)\cdot \mu_{A_{\mathrm{L}}^k}(\mathbf{v})\right]\right]^2}, \tag{8.89}$$

Since t-conorm is nondecreasing, we obtain

$$\left.\frac{\partial \underline{\overline{z}}_j}{\partial \mu_{A_{\mathrm{L}}^l}(\mathbf{v})}\right|_{l:\,\overline{z}_j^l=-1} \leq 0, \tag{8.90}$$

In the second step, for (8.80) and l such that $\overline{z}_j^l = -1$ we obtain

$$\left.\frac{\partial \overline{\overline{z}}_j}{\partial \mu_{A_{\mathrm{U}}^l}(\mathbf{v})}\right|_{l:\,\overline{z}_j^l=-1} = \left.\frac{\partial}{\partial \mu_{A_{\mathrm{U}}^l}(\mathbf{v})} \frac{\sum_{r=1}^{N} \overline{z}_j^r \cdot \overset{N}{\underset{k=1}{S}}\left[\left(\overline{z}_j^k \,\overline{*}\, \overline{z}_j^r\right)\cdot \mu_{A_{\mathrm{U}}^k}(\mathbf{v})\right]}{\sum_{r=1}^{N} \overset{N}{\underset{k=1}{S}}\left[\left(\overline{z}_j^k \,\overline{*}\, \overline{z}_j^r\right)\cdot \mu_{A_{\mathrm{U}}^k}(\mathbf{v})\right]}\right|_{l:\,\overline{z}_j^l=-1}, \tag{8.91}$$

$$\left.\frac{\partial \overline{\overline{z}}_j}{\partial \mu_{A_{\mathrm{U}}^l}(\mathbf{v})}\right|_{l:\,\overline{z}_j^l=-1} = \frac{\left(\begin{array}{l} \left.\frac{\partial \sum_{r=1}^{N} \overline{z}_j^r \cdot \overset{N}{\underset{k=1}{S}}\left[\left(\overline{z}_j^k \,\overline{*}\, \overline{z}_j^r\right)\cdot \mu_{A_{\mathrm{U}}^k}(\mathbf{v})\right]}{\partial \mu_{A_{\mathrm{U}}^l}(\mathbf{v})}\right|_{l:\,\overline{z}_j^l=-1} \cdot \sum_{r=1}^{N} \overset{N}{\underset{k=1}{S}}\left[\left(\overline{z}_j^k \,\overline{*}\, \overline{z}_j^r\right)\cdot \mu_{A_{\mathrm{U}}^k}(\mathbf{v})\right] - \\ \sum_{r=1}^{N} \overline{z}_j^r \cdot \overset{N}{\underset{k=1}{S}}\left[\left(\overline{z}_j^k \,\overline{*}\, \overline{z}_j^r\right)\cdot \mu_{A_{\mathrm{U}}^k}(\mathbf{v})\right] \cdot \left.\frac{\partial \sum_{r=1}^{N} \overset{N}{\underset{k=1}{S}}\left[\left(\overline{z}_j^k \,\overline{*}\, \overline{z}_j^r\right)\cdot \mu_{A_{\mathrm{U}}^k}(\mathbf{v})\right]}{\partial \mu_{A_{\mathrm{U}}^l}(\mathbf{v})}\right|_{l:\,\overline{z}_j^l=-1} \end{array}\right)}{\left[\sum_{r=1}^{N} \overset{N}{\underset{k=1}{S}}\left[\left(\overline{z}_j^k \,\overline{*}\, \overline{z}_j^r\right)\cdot \mu_{A_{\mathrm{U}}^k}(\mathbf{v})\right]\right]^2}, \tag{8.92}$$

$$\left.\frac{\partial \overline{\overline{z}}_j}{\partial \mu_{A_{\mathrm{U}}^l}(\mathbf{v})}\right|_{l:\,\overline{z}_j^l=-1} = \frac{\left(\begin{array}{l} \sum_{r=1}^{N} \overline{z}_j^r \cdot \left.\frac{\partial \overset{N}{\underset{k=1}{S}}\left[\left(\overline{z}_j^k \,\overline{*}\, \overline{z}_j^r\right)\cdot \mu_{A_{\mathrm{U}}^k}(\mathbf{v})\right]}{\partial \mu_{A_{\mathrm{U}}^l}(\mathbf{v})}\right|_{l:\,\overline{z}_j^l=-1} \cdot \sum_{r=1}^{N} \overset{N}{\underset{k=1}{S}}\left[\left(\overline{z}_j^k \,\overline{*}\, \overline{z}_j^r\right)\cdot \mu_{A_{\mathrm{U}}^k}(\mathbf{v})\right] - \\ \sum_{r=1}^{N} \overline{z}_j^r \cdot \overset{N}{\underset{k=1}{S}}\left[\left(\overline{z}_j^k \,\overline{*}\, \overline{z}_j^r\right)\cdot \mu_{A_{\mathrm{U}}^k}(\mathbf{v})\right] \cdot \sum_{r=1}^{N} \left.\frac{\partial \overset{N}{\underset{k=1}{S}}\left[\left(\overline{z}_j^k \,\overline{*}\, \overline{z}_j^r\right)\cdot \mu_{A_{\mathrm{U}}^k}(\mathbf{v})\right]}{\partial \mu_{A_{\mathrm{U}}^l}(\mathbf{v})}\right|_{l:\,\overline{z}_j^l=-1} \end{array}\right)}{\left[\sum_{r=1}^{N} \overset{N}{\underset{k=1}{S}}\left[\left(\overline{z}_j^k \,\overline{*}\, \overline{z}_j^r\right)\cdot \mu_{A_{\mathrm{U}}^k}(\mathbf{v})\right]\right]^2}, \tag{8.93}$$

On the basis of boundary conditions of t-conorms, we get

$$\left.\frac{\partial \overline{z}_j}{\partial \mu_{A_{\mathrm{U}}^l}(\mathbf{v})}\right|_{l:\,\overline{z}_j^l=-1} = \frac{\begin{array}{c}\left(\left(\sum\limits_{\substack{r=1\\ r:\,\overline{z}_j^r=1}}^{N} \left.\frac{\partial \underset{\substack{k=1\\ k:\,\overline{z}_j^k=1}}{\overset{N}{S}}\left[\mu_{A_{\mathrm{U}}^k}(\mathbf{v})\right]}{\partial \mu_{A_{\mathrm{U}}^l}(\mathbf{v})}\right|_{l:\,\overline{z}_j^l=-1} - \sum\limits_{\substack{r=1\\ r:\,\overline{z}_j^r=-1}}^{N} \left.\frac{\partial \underset{\substack{k=1\\ :\,\overline{z}_j^k=-1}}{\overset{N}{S}}\left[\mu_{A_{\mathrm{U}}^k}(\mathbf{v})\right]}{\partial \mu_{A_{\mathrm{U}}^l}(\mathbf{v})}\right|_{l:\,\overline{z}_j^l=-1}\right)\cdot\right.\\ \cdot \sum\limits_{r=1}^{N} \overset{N}{\underset{k=1}{S}}\left[\left(\overline{z}_j^k \overset{=}{*} \overline{z}_j^r\right)\cdot \mu_{A_{\mathrm{U}}^k}(\mathbf{v})\right] - \sum\limits_{r=1}^{N} \overline{z}_j^r \cdot \overset{N}{\underset{k=1}{S}}\left[\left(\overline{z}_j^k \overset{=}{*} \overline{z}_j^r\right)\cdot \mu_{A_{\mathrm{U}}^k}(\mathbf{v})\right]\cdot\\ \left.\cdot\left(\sum\limits_{\substack{r=1\\ \overline{z}_j^r=1}}^{N} \left.\frac{\partial \underset{\substack{k=1\\ \overline{z}_j^k=1}}{\overset{N}{S}}\left[\mu_{A_{\mathrm{U}}^k}(\mathbf{v})\right]}{\partial \mu_{A_{\mathrm{U}}^l}(\mathbf{v})}\right|_{l:\,\overline{z}_j^l=-1} + \sum\limits_{\substack{r=1\\ \overline{z}_j^r=-1}}^{N} \left.\frac{\partial \underset{\substack{k=1\\ \overline{z}_j^k=-1}}{\overset{N}{S}}\left[\mu_{A_{\mathrm{U}}^k}(\mathbf{v})\right]}{\partial \mu_{A_{\mathrm{U}}^l}(\mathbf{v})}\right|_{l:\,\overline{z}_j^l=-1}\right)\right)\end{array}}{\left[\sum\limits_{r=1}^{N} \overset{N}{\underset{k=1}{S}}\left[\left(\overline{z}_j^k \overset{=}{*} \overline{z}_j^r\right)\cdot \mu_{A_{\mathrm{U}}^k}(\mathbf{v})\right]\right]^2}, \tag{8.94}$$

$$\left.\frac{\partial \overline{z}_j}{\partial \mu_{A_{\mathrm{U}}^l}(\mathbf{v})}\right|_{l:\,\overline{z}_j^l=-1} = \frac{\begin{array}{c}\left(\left(0 - \sum\limits_{\substack{r=1\\ r:\,\overline{z}_j^r=-1}}^{N} \left.\frac{\partial \underset{\substack{k=1\\ :\,\overline{z}_j^k=-1}}{\overset{N}{S}}\left[\mu_{A_{\mathrm{U}}^k}(\mathbf{v})\right]}{\partial \mu_{A_{\mathrm{U}}^l}(\mathbf{v})}\right|_{l:\,\overline{z}_j^l=-1}\right)\cdot\right.\\ \cdot \sum\limits_{r=1}^{N} \overset{N}{\underset{k=1}{S}}\left[\left(\overline{z}_j^k \overset{=}{*} \overline{z}_j^r\right)\cdot \mu_{A_{\mathrm{U}}^k}(\mathbf{v})\right] - \sum\limits_{r=1}^{N} \overline{z}_j^r \cdot \overset{N}{\underset{k=1}{S}}\left[\left(\overline{z}_j^k \overset{=}{*} \overline{z}_j^r\right)\cdot \mu_{A_{\mathrm{U}}^k}(\mathbf{v})\right]\cdot\\ \left.\cdot\left(0 + \sum\limits_{\substack{r=1\\ \overline{z}_j^r=-1}}^{N} \left.\frac{\partial \underset{\substack{k=1\\ \overline{z}_j^k=-1}}{\overset{N}{S}}\left[\mu_{A_{\mathrm{U}}^k}(\mathbf{v})\right]}{\partial \mu_{A_{\mathrm{U}}^l}(\mathbf{v})}\right|_{l:\,\overline{z}_j^l=-1}\right)\right)\end{array}}{\left[\sum\limits_{r=1}^{N} \overset{N}{\underset{k=1}{S}}\left[\left(\overline{z}_j^k \overset{=}{*} \overline{z}_j^r\right)\cdot \mu_{A_{\mathrm{U}}^k}(\mathbf{v})\right]\right]^2}, \tag{8.95}$$

$$\left.\frac{\partial \overline{z}_j}{\partial \mu_{A_{\mathrm{U}}^l}(\mathbf{v})}\right|_{l:\,\overline{z}_j^l=-1} = \frac{\begin{array}{c}\left(- \sum\limits_{\substack{r=1\\ r:\,\overline{z}_j^r=-1}}^{N} \left.\frac{\partial \underset{\substack{k=1\\ :\,\overline{z}_j^k=-1}}{\overset{N}{S}}\left[\mu_{A_{\mathrm{U}}^k}(\mathbf{v})\right]}{\partial \mu_{A_{\mathrm{U}}^l}(\mathbf{v})}\right|_{l:\,\overline{z}_j^l=-1} \cdot \sum\limits_{r=1}^{N} \overset{N}{\underset{k=1}{S}}\left[\left(\overline{z}_j^k \overset{=}{*} \overline{z}_j^r\right)\cdot \mu_{A_{\mathrm{U}}^k}(\mathbf{v})\right]\right.\\ \left.- \sum\limits_{r=1}^{N} \overline{z}_j^r \cdot \overset{N}{\underset{k=1}{S}}\left[\left(\overline{z}_j^k \overset{=}{*} \overline{z}_j^r\right)\cdot \mu_{A_{\mathrm{U}}^k}(\mathbf{v})\right]\cdot \sum\limits_{\substack{r=1\\ \overline{z}_j^r=-1}}^{N} \left.\frac{\partial \underset{\substack{k=1\\ \overline{z}_j^k=-1}}{\overset{N}{S}}\left[\mu_{A_{\mathrm{U}}^k}(\mathbf{v})\right]}{\partial \mu_{A_{\mathrm{U}}^l}(\mathbf{v})}\right|_{l:\,\overline{z}_j^l=-1}\right)\end{array}}{\left[\sum\limits_{r=1}^{N} \overset{N}{\underset{k=1}{S}}\left[\left(\overline{z}_j^k \overset{=}{*} \overline{z}_j^r\right)\cdot \mu_{A_{\mathrm{U}}^k}(\mathbf{v})\right]\right]^2}, \tag{8.96}$$

$$\left.\frac{\partial \overline{z}_j}{\partial \mu_{A_{\mathrm{U}}^l}(\mathbf{v})}\right|_{l:\,\overline{z}_j^l=-1} = -\frac{\left(\begin{array}{l}\displaystyle\sum_{\substack{r=1\\ r:\,\overline{z}_j^r=-1}}^{N} \left.\frac{\partial \mathop{S}\limits_{\substack{k=1\\ :\,\overline{z}_j^k=-1}}^{N}\left[\mu_{A_{\mathrm{U}}^k}(\mathbf{v})\right]}{\partial \mu_{A_{\mathrm{U}}^l}(\mathbf{v})}\right|_{l:\,\overline{z}_j^l=-1} \cdot \\ \displaystyle\cdot \sum_{r=1}^{N}\left(\overline{z}_j^r+1\right)\mathop{S}\limits_{k=1}^{N}\left[\left(\overline{z}_j^k \mathbin{\overline{\overline{*}}} \overline{z}_j^r\right)\cdot \mu_{A_{\mathrm{U}}^k}(\mathbf{v})\right]\end{array}\right)}{\left[\displaystyle\sum_{r=1}^{N}\mathop{S}\limits_{k=1}^{N}\left[\left(\overline{z}_j^k \mathbin{\overline{\overline{*}}} \overline{z}_j^r\right)\cdot \mu_{A_{\mathrm{U}}^k}(\mathbf{v})\right]\right]^2}, \tag{8.97}$$

Since t-conorm is nondecreasing, we obtain

$$\left.\frac{\partial \overline{z}_j}{\partial \mu_{A_{\mathrm{U}}^l}(\mathbf{v})}\right|_{l:\,\overline{z}_j^l=-1} \leq 0, \tag{8.98}$$

In the third step for (8.79) and l such that $\overline{z}_j^l = 1$

$$\left.\frac{\partial \underline{z}_j}{\partial \mu_{A_{\mathrm{L}}^l}(\mathbf{v})}\right|_{l:\,\overline{z}_j^l=1} = \left.\frac{\partial}{\partial \mu_{A_{\mathrm{L}}^l}(\mathbf{v})} \frac{\displaystyle\sum_{r=1}^{N}\overline{z}_j^r \cdot \mathop{S}\limits_{k=1}^{N}\left[\left(\overline{z}_j^k \mathbin{\overline{\overline{*}}} \overline{z}_j^r\right)\cdot \mu_{A_{\mathrm{L}}^k}(\mathbf{v})\right]}{\displaystyle\sum_{r=1}^{N}\mathop{S}\limits_{k=1}^{N}\left[\left(\overline{z}_j^k \mathbin{\overline{\overline{*}}} \overline{z}_j^r\right)\cdot \mu_{A_{\mathrm{L}}^k}(\mathbf{v})\right]}\right|_{l:\,\overline{z}_j^l=1}, \tag{8.99}$$

$$\left.\frac{\partial \underline{z}_j}{\partial \mu_{A_{\mathrm{L}}^l}(\mathbf{v})}\right|_{l:\,\overline{z}_j^l=1} = \frac{\left(\begin{array}{l}\left.\dfrac{\partial \sum_{r=1}^{N}\overline{z}_j^r \cdot \mathop{S}\limits_{k=1}^{N}\left[\left(\overline{z}_j^k \mathbin{\overline{\overline{*}}} \overline{z}_j^r\right)\cdot \mu_{A_{\mathrm{L}}^k}(\mathbf{v})\right]}{\partial \mu_{A_{\mathrm{L}}^l}(\mathbf{v})}\right|_{l:\,\overline{z}_j^l=1} \cdot \displaystyle\sum_{r=1}^{N}\mathop{S}\limits_{k=1}^{N}\left[\left(\overline{z}_j^k \mathbin{\overline{\overline{*}}} \overline{z}_j^r\right)\cdot \mu_{A_{\mathrm{L}}^k}(\mathbf{v})\right] - \\ \displaystyle\sum_{r=1}^{N}\overline{z}_j^r \cdot \mathop{S}\limits_{k=1}^{N}\left[\left(\overline{z}_j^k \mathbin{\overline{\overline{*}}} \overline{z}_j^r\right)\cdot \mu_{A_{\mathrm{L}}^k}(\mathbf{v})\right] \cdot \left.\dfrac{\partial \sum_{r=1}^{N}\mathop{S}\limits_{k=1}^{N}\left[\left(\overline{z}_j^k \mathbin{\overline{\overline{*}}} \overline{z}_j^r\right)\cdot \mu_{A_{\mathrm{L}}^k}(\mathbf{v})\right]}{\partial \mu_{A_{\mathrm{L}}^l}(\mathbf{v})}\right|_{l:\,\overline{z}_j^l=1}\end{array}\right)}{\left[\displaystyle\sum_{r=1}^{N}\mathop{S}\limits_{k=1}^{N}\left[\left(\overline{z}_j^k \mathbin{\overline{\overline{*}}} \overline{z}_j^r\right)\cdot \mu_{A_{\mathrm{L}}^k}(\mathbf{v})\right]\right]^2}, \tag{8.100}$$

$$\left.\frac{\partial \underline{z}_j}{\partial \mu_{A_{\mathrm{L}}^l}(\mathbf{v})}\right|_{l:\,\overline{z}_j^l=1} = \frac{\left(\begin{array}{l}\displaystyle\sum_{r=1}^{N}\overline{z}_j^r \cdot \left.\dfrac{\partial \mathop{S}\limits_{k=1}^{N}\left[\left(\overline{z}_j^k \mathbin{\overline{\overline{*}}} \overline{z}_j^r\right)\cdot \mu_{A_{\mathrm{L}}^k}(\mathbf{v})\right]}{\partial \mu_{A_{\mathrm{L}}^l}(\mathbf{v})}\right|_{l:\,\overline{z}_j^l=1} \cdot \displaystyle\sum_{r=1}^{N}\mathop{S}\limits_{k=1}^{N}\left[\left(\overline{z}_j^k \mathbin{\overline{\overline{*}}} \overline{z}_j^r\right)\cdot \mu_{A_{\mathrm{L}}^k}(\mathbf{v})\right] - \\ \displaystyle\sum_{r=1}^{N}\overline{z}_j^r \cdot \mathop{S}\limits_{k=1}^{N}\left[\left(\overline{z}_j^k \mathbin{\overline{\overline{*}}} \overline{z}_j^r\right)\cdot \mu_{A_{\mathrm{L}}^k}(\mathbf{v})\right] \cdot \sum_{r=1}^{N} \left.\dfrac{\partial \mathop{S}\limits_{k=1}^{N}\left[\left(\overline{z}_j^k \mathbin{\overline{\overline{*}}} \overline{z}_j^r\right)\cdot \mu_{A_{\mathrm{L}}^k}(\mathbf{v})\right]}{\partial \mu_{A_{\mathrm{L}}^l}(\mathbf{v})}\right|_{l:\,\overline{z}_j^l=1}\end{array}\right)}{\left[\displaystyle\sum_{r=1}^{N}\mathop{S}\limits_{k=1}^{N}\left[\left(\overline{z}_j^k \mathbin{\overline{\overline{*}}} \overline{z}_j^r\right)\cdot \mu_{A_{\mathrm{L}}^k}(\mathbf{v})\right]\right]^2}, \tag{8.101}$$

On the basis of boundary conditions of t-conorms, we get

$$\left.\frac{\partial \underline{\overline{z}}_j}{\partial \mu_{A_{\mathrm{L}}^l}(\mathbf{v})}\right|_{l:\,\overline{\overline{z}}_j^l=1} = \frac{\left(\begin{array}{l} \left(\sum\limits_{\substack{r=1 \\ r:\,\overline{\overline{z}}_j^r=1}}^{N} \left.\dfrac{\partial \underset{\substack{k=1 \\ k:\,\overline{\overline{z}}_j^k=1}}{\overset{N}{S}} \left[\mu_{A_{\mathrm{L}}^k}(\mathbf{v})\right]}{\partial \mu_{A_{\mathrm{L}}^l}(\mathbf{v})}\right|_{l:\,\overline{\overline{z}}_j^l=1} - \sum\limits_{\substack{r=1 \\ r:\,\overline{\overline{z}}_j^r=-1}}^{N} \left.\dfrac{\partial \underset{\substack{k=1 \\ :\,\overline{\overline{z}}_j^k=-1}}{\overset{N}{S}} \left[\mu_{A_{\mathrm{L}}^k}(\mathbf{v})\right]}{\partial \mu_{A_{\mathrm{L}}^l}(\mathbf{v})}\right|_{l:\,\overline{\overline{z}}_j^l=1} \right) \cdot \\ \cdot \sum\limits_{r=1}^{N} \underset{k=1}{\overset{N}{S}} \left[\left(\overline{\overline{z}}_j^k \mathbin{\overline{\overline{*}}} \overline{\overline{z}}_j^r\right) \cdot \mu_{A_{\mathrm{L}}^k}(\mathbf{v})\right] - \sum\limits_{r=1}^{N} \overline{\overline{z}}_j^r \cdot \underset{k=1}{\overset{N}{S}} \left[\left(\overline{\overline{z}}_j^k \mathbin{\overline{\overline{*}}} \overline{\overline{z}}_j^r\right) \cdot \mu_{A_{\mathrm{L}}^k}(\mathbf{v})\right] \cdot \\ \cdot \left(\sum\limits_{\substack{r=1 \\ \overline{\overline{z}}_j^r=1}}^{N} \left.\dfrac{\partial \underset{\substack{k=1 \\ \overline{\overline{z}}_j^k=1}}{\overset{N}{S}} \left[\mu_{A_{\mathrm{L}}^k}(\mathbf{v})\right]}{\partial \mu_{A_{\mathrm{L}}^l}(\mathbf{v})}\right|_{l:\,\overline{\overline{z}}_j^l=1} + \sum\limits_{\substack{r=1 \\ \overline{\overline{z}}_j^r=-1}}^{N} \left.\dfrac{\partial \underset{\substack{k=1 \\ \overline{\overline{z}}_j^k=-1}}{\overset{N}{S}} \left[\mu_{A_{\mathrm{L}}^k}(\mathbf{v})\right]}{\partial \mu_{A_{\mathrm{L}}^l}(\mathbf{v})}\right|_{l:\,\overline{\overline{z}}_j^l=1} \right) \end{array}\right)}{\left[\sum\limits_{r=1}^{N} \underset{k=1}{\overset{N}{S}} \left[\left(\overline{\overline{z}}_j^k \mathbin{\overline{\overline{*}}} \overline{\overline{z}}_j^r\right) \cdot \mu_{A_{\mathrm{L}}^k}(\mathbf{v})\right]\right]^2}, \tag{8.102}$$

$$\left.\frac{\partial \underline{\overline{z}}_j}{\partial \mu_{A_{\mathrm{L}}^l}(\mathbf{v})}\right|_{l:\,\overline{\overline{z}}_j^l=1} = \frac{\left(\begin{array}{l} \left(\sum\limits_{\substack{r=1 \\ r:\,\overline{\overline{z}}_j^r=1}}^{N} \left.\dfrac{\partial \underset{\substack{k=1 \\ k:\,\overline{\overline{z}}_j^k=1}}{\overset{N}{S}} \left[\mu_{A_{\mathrm{L}}^k}(\mathbf{v})\right]}{\partial \mu_{A_{\mathrm{L}}^l}(\mathbf{v})}\right|_{l:\,\overline{\overline{z}}_j^l=1} - 0 \right) \cdot \sum\limits_{r=1}^{N} \underset{k=1}{\overset{N}{S}} \left[\left(\overline{\overline{z}}_j^k \mathbin{\overline{\overline{*}}} \overline{\overline{z}}_j^r\right) \cdot \mu_{A_{\mathrm{L}}^k}(\mathbf{v})\right] \\ - \sum\limits_{r=1}^{N} \overline{\overline{z}}_j^r \cdot \underset{k=1}{\overset{N}{S}} \left[\left(\overline{\overline{z}}_j^k \mathbin{\overline{\overline{*}}} \overline{\overline{z}}_j^r\right) \cdot \mu_{A_{\mathrm{L}}^k}(\mathbf{v})\right] \cdot \left(\sum\limits_{\substack{r=1 \\ \overline{\overline{z}}_j^r=1}}^{N} \left.\dfrac{\partial \underset{\substack{k=1 \\ \overline{\overline{z}}_j^k=1}}{\overset{N}{S}} \left[\mu_{A_{\mathrm{L}}^k}(\mathbf{v})\right]}{\partial \mu_{A_{\mathrm{L}}^l}(\mathbf{v})}\right|_{l:\,\overline{\overline{z}}_j^l=1} + 0 \right) \end{array}\right)}{\left[\sum\limits_{r=1}^{N} \underset{k=1}{\overset{N}{S}} \left[\left(\overline{\overline{z}}_j^k \mathbin{\overline{\overline{*}}} \overline{\overline{z}}_j^r\right) \cdot \mu_{A_{\mathrm{L}}^k}(\mathbf{v})\right]\right]^2}, \tag{8.103}$$

$$\left.\frac{\partial \underline{\overline{z}}_j}{\partial \mu_{A_{\mathrm{L}}^l}(\mathbf{v})}\right|_{l:\,\overline{\overline{z}}_j^l=1} = \frac{\left(\sum\limits_{\substack{r=1 \\ r:\,\overline{\overline{z}}_j^r=1}}^{N} \left.\dfrac{\partial \underset{\substack{k=1 \\ k:\,\overline{\overline{z}}_j^k=1}}{\overset{N}{S}} \left[\mu_{A_{\mathrm{L}}^k}(\mathbf{v})\right]}{\partial \mu_{A_{\mathrm{L}}^l}(\mathbf{v})}\right|_{l:\,\overline{\overline{z}}_j^l=1} \cdot \sum\limits_{r=1}^{N} \left(1 - \overline{\overline{z}}_j^r\right) \cdot \underset{k=1}{\overset{N}{S}} \left[\left(\overline{\overline{z}}_j^k \mathbin{\overline{\overline{*}}} \overline{\overline{z}}_j^r\right) \cdot \mu_{A_{\mathrm{L}}^k}(\mathbf{v})\right] \right)}{\left[\sum\limits_{r=1}^{N} \underset{k=1}{\overset{N}{S}} \left[\left(\overline{\overline{z}}_j^k \mathbin{\overline{\overline{*}}} \overline{\overline{z}}_j^r\right) \cdot \mu_{A_{\mathrm{L}}^k}(\mathbf{v})\right]\right]^2}, \tag{8.104}$$

Since t-conorm is nondecreasing, we obtain

$$\left.\frac{\partial \underline{\overline{z}}_j}{\partial \mu_{A_{\mathrm{L}}^l}(\mathbf{v})}\right|_{l:\,\overline{\overline{z}}_j^l=1} \geq 0, \tag{8.105}$$

In the fourth step for (8.80) and l such that $\overline{z}_j^l = 1$

$$\left.\frac{\partial \overline{\overline{z}}_j}{\partial \mu_{A_{\mathrm{U}}^l}(\mathbf{v})}\right|_{l:\,\overline{z}_j^l=1} = \frac{\partial}{\partial \mu_{A_{\mathrm{U}}^l}(\mathbf{v})} \left.\frac{\sum_{r=1}^{N} \overline{z}_j^r \cdot \mathop{S}_{k=1}^{N}\left[\left(\overline{z}_j^k \overset{=}{*} \overline{z}_j^r\right) \cdot \mu_{A_{\mathrm{U}}^k}(\mathbf{v})\right]}{\sum_{r=1}^{N} \mathop{S}_{k=1}^{N}\left[\left(\overline{z}_j^k \overset{=}{*} \overline{z}_j^r\right) \cdot \mu_{A_{\mathrm{U}}^k}(\mathbf{v})\right]}\right|_{l:\,\overline{z}_j^l=1}, \tag{8.106}$$

$$\left.\frac{\partial \overline{\overline{z}}_j}{\partial \mu_{A_{\mathrm{U}}^l}(\mathbf{v})}\right|_{l:\,\overline{z}_j^l=1} = \frac{\left(\begin{array}{l} \left.\frac{\partial \sum_{r=1}^{N} \overline{z}_j^r \cdot \mathop{S}_{k=1}^{N}\left[\left(\overline{z}_j^k \overset{=}{*} \overline{z}_j^r\right) \cdot \mu_{A_{\mathrm{U}}^k}(\mathbf{v})\right]}{\partial \mu_{A_{\mathrm{U}}^l}(\mathbf{v})}\right|_{l:\,\overline{z}_j^l=1} \cdot \sum_{r=1}^{N} \mathop{S}_{k=1}^{N}\left[\left(\overline{z}_j^k \overset{=}{*} \overline{z}_j^r\right) \cdot \mu_{A_{\mathrm{U}}^k}(\mathbf{v})\right] - \\ \sum_{r=1}^{N} \overline{z}_j^r \cdot \mathop{S}_{k=1}^{N}\left[\left(\overline{z}_j^k \overset{=}{*} \overline{z}_j^r\right) \cdot \mu_{A_{\mathrm{U}}^k}(\mathbf{v})\right] \cdot \left.\frac{\partial \sum_{r=1}^{N} \mathop{S}_{k=1}^{N}\left[\left(\overline{z}_j^k \overset{=}{*} \overline{z}_j^r\right) \cdot \mu_{A_{\mathrm{U}}^k}(\mathbf{v})\right]}{\partial \mu_{A_{\mathrm{U}}^l}(\mathbf{v})}\right|_{l:\,\overline{z}_j^l=1} \end{array}\right)}{\left[\sum_{r=1}^{N} \mathop{S}_{k=1}^{N}\left[\left(\overline{z}_j^k \overset{=}{*} \overline{z}_j^r\right) \cdot \mu_{A_{\mathrm{U}}^k}(\mathbf{v})\right]\right]^2}, \tag{8.107}$$

$$\left.\frac{\partial \overline{\overline{z}}_j}{\partial \mu_{A_{\mathrm{U}}^l}(\mathbf{v})}\right|_{l:\,\overline{z}_j^l=1} = \frac{\left(\begin{array}{l} \sum_{r=1}^{N} \overline{z}_j^r \cdot \left.\frac{\partial \mathop{S}_{k=1}^{N}\left[\left(\overline{z}_j^k \overset{=}{*} \overline{z}_j^r\right) \cdot \mu_{A_{\mathrm{U}}^k}(\mathbf{v})\right]}{\partial \mu_{A_{\mathrm{U}}^l}(\mathbf{v})}\right|_{l:\,\overline{z}_j^l=1} \cdot \sum_{r=1}^{N} \mathop{S}_{k=1}^{N}\left[\left(\overline{z}_j^k \overset{=}{*} \overline{z}_j^r\right) \cdot \mu_{A_{\mathrm{U}}^k}(\mathbf{v})\right] - \\ \sum_{r=1}^{N} \overline{z}_j^r \cdot \mathop{S}_{k=1}^{N}\left[\left(\overline{z}_j^k \overset{=}{*} \overline{z}_j^r\right) \cdot \mu_{A_{\mathrm{U}}^k}(\mathbf{v})\right] \cdot \sum_{r=1}^{N} \left.\frac{\partial \mathop{S}_{k=1}^{N}\left[\left(\overline{z}_j^k \overset{=}{*} \overline{z}_j^r\right) \cdot \mu_{A_{\mathrm{U}}^k}(\mathbf{v})\right]}{\partial \mu_{A_{\mathrm{U}}^l}(\mathbf{v})}\right|_{l:\,\overline{z}_j^l=1} \end{array}\right)}{\left[\sum_{r=1}^{N} \mathop{S}_{k=1}^{N}\left[\left(\overline{z}_j^k \overset{=}{*} \overline{z}_j^r\right) \cdot \mu_{A_{\mathrm{U}}^k}(\mathbf{v})\right]\right]^2}, \tag{8.108}$$

On the basis of boundary conditions of t-conorms, we get

$$\left.\frac{\partial \overline{\overline{z}}_j}{\partial \mu_{A_{\mathrm{U}}^l}(\mathbf{v})}\right|_{l:\,\overline{z}_j^l=1} = \frac{\left(\begin{array}{l} \left(\sum_{\substack{r=1 \\ r:\,\overline{z}_j^r=1}}^{N} \left.\frac{\partial \mathop{S}_{\substack{k=1 \\ k:\,\overline{z}_j^k=1}}^{N}\left[\mu_{A_{\mathrm{U}}^k}(\mathbf{v})\right]}{\partial \mu_{A_{\mathrm{U}}^l}(\mathbf{v})}\right|_{l:\,\overline{z}_j^l=1} - \sum_{\substack{r=1 \\ r:\,\overline{z}_j^r=-1}}^{N} \left.\frac{\partial \mathop{S}_{\substack{k=1 \\ :\,\overline{z}_j^k=-1}}^{N}\left[\mu_{A_{\mathrm{U}}^k}(\mathbf{v})\right]}{\partial \mu_{A_{\mathrm{U}}^l}(\mathbf{v})}\right|_{l:\,\overline{z}_j^l=1}\right) \cdot \\ \cdot \sum_{r=1}^{N} \mathop{S}_{k=1}^{N}\left[\left(\overline{z}_j^k \overset{=}{*} \overline{z}_j^r\right) \cdot \mu_{A_{\mathrm{U}}^k}(\mathbf{v})\right] - \sum_{r=1}^{N} \overline{z}_j^r \cdot \mathop{S}_{k=1}^{N}\left[\left(\overline{z}_j^k \overset{=}{*} \overline{z}_j^r\right) \cdot \mu_{A_{\mathrm{U}}^k}(\mathbf{v})\right] \cdot \\ \cdot \left(\sum_{\substack{r=1 \\ \overline{z}_j^r=1}}^{N} \left.\frac{\partial \mathop{S}_{\substack{k=1 \\ \overline{z}_j^k=1}}^{N}\left[\mu_{A_{\mathrm{U}}^k}(\mathbf{v})\right]}{\partial \mu_{A_{\mathrm{U}}^l}(\mathbf{v})}\right|_{l:\,\overline{z}_j^l=1} + \sum_{\substack{r=1 \\ \overline{z}_j^r=-1}}^{N} \left.\frac{\partial \mathop{S}_{\substack{k=1 \\ \overline{z}_j^k=-1}}^{N}\left[\mu_{A_{\mathrm{U}}^k}(\mathbf{v})\right]}{\partial \mu_{A_{\mathrm{U}}^l}(\mathbf{v})}\right|_{l:\,\overline{z}_j^l=1}\right) \end{array}\right)}{\left[\sum_{r=1}^{N} \mathop{S}_{k=1}^{N}\left[\left(\overline{z}_j^k \overset{=}{*} \overline{z}_j^r\right) \cdot \mu_{A_{\mathrm{U}}^k}(\mathbf{v})\right]\right]^2}, \tag{8.109}$$

$$\left.\frac{\partial\overline{z}_j}{\partial\mu_{A_{\mathrm{U}}^l}(\mathbf{v})}\right|_{l:\,\overline{z}_j^l=1}=\frac{\left(\begin{array}{c}\left(\displaystyle\sum_{\substack{r=1\\ r:\,\overline{z}_j^r=1}}^{N}\left.\frac{\partial\displaystyle\mathop{S}_{\substack{k=1\\ k:\,\overline{z}_j^k=1}}^{N}\left[\mu_{A_{\mathrm{U}}^k}(\mathbf{v})\right]}{\partial\mu_{A_{\mathrm{U}}^l}(\mathbf{v})}\right|_{l:\,\overline{z}_j^l=1}-0\right)\cdot\displaystyle\sum_{r=1}^{N}\mathop{S}_{k=1}^{N}\left[\left(\overline{z}_j^k\overset{=}{*}\overline{z}_j^r\right)\cdot\mu_{A_{\mathrm{U}}^k}(\mathbf{v})\right]\\ -\displaystyle\sum_{r=1}^{N}\overline{z}_j^r\cdot\mathop{S}_{k=1}^{N}\left[\left(\overline{z}_j^k\overset{=}{*}\overline{z}_j^r\right)\cdot\mu_{A_{\mathrm{U}}^k}(\mathbf{v})\right]\cdot\left(\displaystyle\sum_{\substack{r=1\\ \overline{z}_j^r=1}}^{N}\left.\frac{\partial\displaystyle\mathop{S}_{\substack{k=1\\ \overline{z}_j^k=1}}^{N}\left[\mu_{A_{\mathrm{U}}^k}(\mathbf{v})\right]}{\partial\mu_{A_{\mathrm{U}}^l}(\mathbf{v})}\right|_{l:\,\overline{z}_j^l=1}+0\right)\end{array}\right)}{\left[\displaystyle\sum_{r=1}^{N}\mathop{S}_{k=1}^{N}\left[\left(\overline{z}_j^k\overset{=}{*}\overline{z}_j^r\right)\cdot\mu_{A_{\mathrm{U}}^k}(\mathbf{v})\right]\right]^2},\tag{8.110}$$

$$\left.\frac{\partial\overline{z}_j}{\partial\mu_{A_{\mathrm{U}}^l}(\mathbf{v})}\right|_{l:\,\overline{z}_j^l=1}=\frac{\left(\displaystyle\sum_{\substack{r=1\\ r:\,\overline{z}_j^r=1}}^{N}\left.\frac{\partial\displaystyle\mathop{S}_{\substack{k=1\\ k:\,\overline{z}_j^k=1}}^{N}\left[\mu_{A_{\mathrm{U}}^k}(\mathbf{v})\right]}{\partial\mu_{A_{\mathrm{U}}^l}(\mathbf{v})}\right|_{l:\,\overline{z}_j^l=1}\cdot\displaystyle\sum_{r=1}^{N}\left(1-\overline{z}_j^r\right)\cdot\mathop{S}_{k=1}^{N}\left[\left(\overline{z}_j^k\overset{=}{*}\overline{z}_j^r\right)\cdot\mu_{A_{\mathrm{U}}^k}(\mathbf{v})\right]\right)}{\left[\displaystyle\sum_{r=1}^{N}\mathop{S}_{k=1}^{N}\left[\left(\overline{z}_j^k\overset{=}{*}\overline{z}_j^r\right)\cdot\mu_{A_{\mathrm{U}}^k}(\mathbf{v})\right]\right]^2},\tag{8.111}$$

Since t-conorm is nondecreasing, we obtain

$$\left.\frac{\partial\overline{z}_j}{\partial\mu_{A_{\mathrm{U}}^l}(\mathbf{v})}\right|_{l:\,\overline{z}_j^l=1}\geq 0,\tag{8.112}$$

Analyzing inequalities (8.90), (8.98), (8.105) and (8.112) it is easily seen that in formulas (8.79) and (8.80) fuzzy sets A_{L}^r and A_{U}^r should be chosen according to descriptions (8.81) and (8.82).

8.3.3 S/D–DCOG Rough Fuzzy Classifier

Following formula (8.43), the lower and upper approximations of the membership of object x to class ω_j are given by

$$\underline{z}_j=\frac{\displaystyle\sum_{r=1}^{N}\overline{z}_j^r\cdot\mathop{T}_{k=1}^{N}\left[N\left(T\left(\mu_{A_{\mathrm{L}}^k}(\mathbf{v}),\overline{z}_j^k\overset{\neq}{*}\overline{z}_j^r\right)\right)\right]}{\displaystyle\sum_{r=1}^{N}\mathop{T}_{k=1}^{N}\left[N\left(T\left(\mu_{A_{\mathrm{L}}^k}(\mathbf{v}),\overline{z}_j^k\overset{\neq}{*}\overline{z}_j^r\right)\right)\right]}.\tag{8.113}$$

$$\overline{z}_j=\frac{\displaystyle\sum_{r=1}^{N}\overline{z}_j^r\cdot\mathop{T}_{k=1}^{N}\left[N\left(T\left(\mu_{A_{\mathrm{U}}^k}(\mathbf{v}),\overline{z}_j^k\overset{\neq}{*}\overline{z}_j^r\right)\right)\right]}{\displaystyle\sum_{r=1}^{N}\mathop{T}_{k=1}^{N}\left[N\left(T\left(\mu_{A_{\mathrm{U}}^k}(\mathbf{v}),\overline{z}_j^k\overset{\neq}{*}\overline{z}_j^r\right)\right)\right]}.\tag{8.114}$$

Theorem 8.3. *Fuzzy sets* A_{L}^r *and* A_{U}^r *in descriptions* (8.113) *and* (8.114) *should be chosen as follows*

$$A_{\mathrm{L}}^r = \begin{cases} \underline{\widetilde{P}}A^r & \text{if } \overline{\mathrm{z}}_{\mathrm{j}}^{\mathrm{r}} = 1 \\ \overline{\widetilde{P}}A^r & \text{if } \overline{\mathrm{z}}_{\mathrm{j}}^{\mathrm{r}} = -1. \end{cases} \tag{8.115}$$

$$A_{\mathrm{U}}^r = \begin{cases} \overline{\widetilde{P}}A^r & \text{if } \overline{\mathrm{z}}_{\mathrm{j}}^{\mathrm{r}} = 1 \\ \underline{\widetilde{P}}A^r & \text{if } \overline{\mathrm{z}}_{\mathrm{j}}^{\mathrm{r}} = -1. \end{cases} \tag{8.116}$$

Proof (Theorem 8.3). In the first step, for (8.113) and $l : \overline{z}_j^l = -1$:

$$\left.\frac{\partial \underline{\overline{z}}_j}{\partial \mu_{A_{\mathrm{L}}^l}(\mathbf{v})}\right|_{l:\,\overline{z}_j^l=-1} = \left.\frac{\partial}{\partial \mu_{A_{\mathrm{L}}^l}(\mathbf{v})} \frac{\sum\limits_{r=1}^{N} \overline{z}_j^r \cdot \mathop{T}\limits_{k=1}^{N} \left[N\left(T\left(\mu_{A_{\mathrm{L}}^k}(\mathbf{v}), \overline{z}_j^k \overset{\neq}{*} \overline{z}_j^r \right)\right)\right]}{\sum\limits_{r=1}^{N} \mathop{T}\limits_{k=1}^{N} \left[N\left(T\left(\mu_{A_{\mathrm{L}}^k}(\mathbf{v}), \overline{z}_j^k \overset{\neq}{*} \overline{z}_j^r \right)\right)\right]}\right|_{l:\,\overline{z}_j^l=-1}, \tag{8.117}$$

$$\left.\frac{\partial \underline{\overline{z}}_j}{\partial \mu_{A_{\mathrm{L}}^l}(\mathbf{v})}\right|_{l:\,\overline{z}_j^l=-1} = \frac{\left(\begin{array}{c} \left.\dfrac{\partial \sum\limits_{r=1}^{N} \overline{z}_j^r \cdot \mathop{T}\limits_{k=1}^{N} \left[N\left(T\left(\mu_{A_{\mathrm{L}}^k}(\mathbf{v}), \overline{z}_j^k \overset{\neq}{*} \overline{z}_j^r \right)\right)\right]}{\partial \mu_{A_{\mathrm{L}}^l}(\mathbf{v})}\right|_{l:\,\overline{z}_j^l=-1} \cdot \\ \sum\limits_{r=1}^{N} \mathop{T}\limits_{k=1}^{N} \left[N\left(T\left(\mu_{A_{\mathrm{L}}^k}(\mathbf{v}), \overline{z}_j^k \overset{\neq}{*} \overline{z}_j^r \right)\right)\right] \\ -\sum\limits_{r=1}^{N} \overline{z}_j^r \cdot \mathop{T}\limits_{k=1}^{N} \left[N\left(T\left(\mu_{A_{\mathrm{L}}^k}(\mathbf{v}), \overline{z}_j^k \overset{\neq}{*} \overline{z}_j^r \right)\right)\right] \cdot \\ \left.\dfrac{\partial \sum\limits_{r=1}^{N} \mathop{T}\limits_{k=1}^{N} \left[N\left(T\left(\mu_{A_{\mathrm{L}}^k}(\mathbf{v}), \overline{z}_j^k \overset{\neq}{*} \overline{z}_j^r \right)\right)\right]}{\partial \mu_{A_{\mathrm{L}}^l}(\mathbf{v})}\right|_{l:\,\overline{z}_j^l=-1} \end{array}\right)}{\left[\sum\limits_{r=1}^{N} \mathop{T}\limits_{k=1}^{N} \left[N\left(T\left(\mu_{A_{\mathrm{L}}^k}(\mathbf{v}), \overline{z}_j^k \overset{\neq}{*} \overline{z}_j^r \right)\right)\right]\right]^2}, \tag{8.118}$$

$$\left.\frac{\partial \underline{\overline{z}}_j}{\partial \mu_{A_{\mathrm{L}}^l}(\mathbf{v})}\right|_{l:\,\overline{z}_j^l=-1} = \frac{\left(\begin{array}{c} \sum\limits_{r=1}^{N} \overline{z}_j^r \left.\dfrac{\partial \mathop{T}\limits_{k=1}^{N} \left[N\left(T\left(\mu_{A_{\mathrm{L}}^k}(\mathbf{v}), \overline{z}_j^k \overset{\neq}{*} \overline{z}_j^r \right)\right)\right]}{\partial \mu_{A_{\mathrm{L}}^l}(\mathbf{v})}\right|_{l:\,\overline{z}_j^l=-1} \cdot \\ \sum\limits_{r=1}^{N} \mathop{T}\limits_{k=1}^{N} \left[N\left(T\left(\mu_{A_{\mathrm{L}}^k}(\mathbf{v}), \overline{z}_j^k \overset{\neq}{*} \overline{z}_j^r \right)\right)\right] \\ -\sum\limits_{r=1}^{N} \overline{z}_j^r \cdot \mathop{T}\limits_{k=1}^{N} \left[N\left(T\left(\mu_{A_{\mathrm{L}}^k}(\mathbf{v}), \overline{z}_j^k \overset{\neq}{*} \overline{z}_j^r \right)\right)\right] \cdot \\ \sum\limits_{r=1}^{N} \left.\dfrac{\partial \mathop{T}\limits_{k=1}^{N} \left[N\left(T\left(\mu_{A_{\mathrm{L}}^k}(\mathbf{v}), \overline{z}_j^k \overset{\neq}{*} \overline{z}_j^r \right)\right)\right]}{\partial \mu_{A_{\mathrm{L}}^l}(\mathbf{v})}\right|_{l:\,\overline{z}_j^l=-1} \end{array}\right)}{\left[\sum\limits_{r=1}^{N} \mathop{T}\limits_{k=1}^{N} \left[N\left(T\left(\mu_{A_{\mathrm{L}}^k}(\mathbf{v}), \overline{z}_j^k \overset{\neq}{*} \overline{z}_j^r \right)\right)\right]\right]^2}, \tag{8.119}$$

On the basis of boundary conditions of t-conorms, we get

$$\left.\frac{\partial \overline{z}_j}{\partial \mu_{A_{\mathrm{L}}^l}(\mathbf{v})}\right|_{l:\,\overline{z}_j^l=-1} = \frac{\left(\begin{array}{c}\left(\sum\limits_{\substack{r=1\\ r:\,\overline{z}_j^r=1}}^{N} \left.\frac{\partial \mathop{T}\limits_{\substack{k=1\\ k:\,\overline{z}_j^k=-1}}^{N}\left[N\left(\mu_{A_{\mathrm{L}}^k}(\mathbf{v})\right)\right]}{\partial \mu_{A_{\mathrm{L}}^l}(\mathbf{v})}\right|_{l:\,\overline{z}_j^l=-1} - \sum\limits_{\substack{r=1\\ r:\,\overline{z}_j^r=-1}}^{N} \left.\frac{\partial \mathop{T}\limits_{\substack{k=1\\ k:\,\overline{z}_j^k=1}}^{N}\left[N\left(\mu_{A_{\mathrm{L}}^k}(\mathbf{v})\right)\right]}{\partial \mu_{A_{\mathrm{L}}^l}(\mathbf{v})}\right|_{l:\,\overline{z}_j^l=-1}\right)\cdot\\ \sum\limits_{r=1}^{N}\mathop{T}\limits_{k=1}^{N}\left[N\left(T\left(\mu_{A_{\mathrm{L}}^k}(\mathbf{v}),\overline{z}_j^k \overset{\neq}{*} \overline{z}_j^r\right)\right)\right]\\ -\sum\limits_{r=1}^{N}\overline{z}_j^r\cdot\mathop{T}\limits_{k=1}^{N}\left[N\left(T\left(\mu_{A_{\mathrm{L}}^k}(\mathbf{v}),\overline{z}_j^k \overset{\neq}{*} \overline{z}_j^r\right)\right)\right]\cdot\\ \left(\sum\limits_{\substack{r=1\\ r:\,\overline{z}_j^r=1}}^{N} \left.\frac{\partial \mathop{T}\limits_{\substack{k=1\\ k:\,\overline{z}_j^k=-1}}^{N}\left[N\left(\mu_{A_{\mathrm{L}}^k}(\mathbf{v})\right)\right]}{\partial \mu_{A_{\mathrm{L}}^l}(\mathbf{v})}\right|_{l:\,\overline{z}_j^l=-1} + \sum\limits_{\substack{r=1\\ r:\,\overline{z}_j^r=-1}}^{N} \left.\frac{\partial \mathop{T}\limits_{\substack{k=1\\ k:\,\overline{z}_j^k=1}}^{N}\left[N\left(\mu_{A_{\mathrm{L}}^k}(\mathbf{v})\right)\right]}{\partial \mu_{A_{\mathrm{L}}^l}(\mathbf{v})}\right|_{l:\,\overline{z}_j^l=-1}\right)\end{array}\right)}{\left[\sum\limits_{r=1}^{N}\mathop{T}\limits_{k=1}^{N}\left[N\left(T\left(\mu_{A_{\mathrm{L}}^k}(\mathbf{v}),\overline{z}_j^k \overset{\neq}{*} \overline{z}_j^r\right)\right)\right]\right]^2}, \tag{8.120}$$

$$\left.\frac{\partial \overline{z}_j}{\partial \mu_{A_{\mathrm{L}}^l}(\mathbf{v})}\right|_{l:\,\overline{z}_j^l=-1} = \frac{\left(\begin{array}{c}\left(\sum\limits_{\substack{r=1\\ r:\,\overline{z}_j^r=1}}^{N} \left.\frac{\partial \mathop{T}\limits_{\substack{k=1\\ k:\,\overline{z}_j^k=-1}}^{N}\left[N\left(\mu_{A_{\mathrm{L}}^k}(\mathbf{v})\right)\right]}{\partial \mu_{A_{\mathrm{L}}^l}(\mathbf{v})}\right|_{l:\,\overline{z}_j^l=-1} - 0\right)\cdot\\ \sum\limits_{r=1}^{N}\mathop{T}\limits_{k=1}^{N}\left[N\left(T\left(\mu_{A_{\mathrm{L}}^k}(\mathbf{v}),\overline{z}_j^k \overset{\neq}{*} \overline{z}_j^r\right)\right)\right]\\ -\sum\limits_{r=1}^{N}\overline{z}_j^r\cdot\mathop{T}\limits_{k=1}^{N}\left[N\left(T\left(\mu_{A_{\mathrm{L}}^k}(\mathbf{v}),\overline{z}_j^k \overset{\neq}{*} \overline{z}_j^r\right)\right)\right]\cdot\\ \left(\sum\limits_{\substack{r=1\\ r:\,\overline{z}_j^r=1}}^{N} \left.\frac{\partial \mathop{T}\limits_{\substack{k=1\\ k:\,\overline{z}_j^k=-1}}^{N}\left[N\left(\mu_{A_{\mathrm{L}}^k}(\mathbf{v})\right)\right]}{\partial \mu_{A_{\mathrm{L}}^l}(\mathbf{v})}\right|_{l:\,\overline{z}_j^l=-1} + 0\right)\end{array}\right)}{\left[\sum\limits_{r=1}^{N}\mathop{T}\limits_{k=1}^{N}\left[N\left(T\left(\mu_{A_{\mathrm{L}}^k}(\mathbf{v}),\overline{z}_j^k \overset{\neq}{*} \overline{z}_j^r\right)\right)\right]\right]^2}, \tag{8.121}$$

$$\left.\frac{\partial \overline{z}_j}{\partial \mu_{A_{\mathrm{L}}^l}(\mathbf{v})}\right|_{l:\,\overline{z}_j^l=-1} = \frac{\left(\begin{array}{c}\sum\limits_{\substack{r=1\\ r:\,\overline{z}_j^r=1}}^{N} \left.\frac{\partial \mathop{T}\limits_{\substack{k=1\\ k:\,\overline{z}_j^k=-1}}^{N}\left[N\left(\mu_{A_{\mathrm{L}}^k}(\mathbf{v})\right)\right]}{\partial \mu_{A_{\mathrm{L}}^l}(\mathbf{v})}\right|_{l:\,\overline{z}_j^l=-1}\cdot\\ \cdot\sum\limits_{r=1}^{N}\left(1-\overline{z}_j^r\right)\cdot\mathop{T}\limits_{k=1}^{N}\left[N\left(T\left(\mu_{A_{\mathrm{L}}^k}(\mathbf{v}),\overline{z}_j^k \overset{\neq}{*} \overline{z}_j^r\right)\right)\right]\end{array}\right)}{\left[\sum\limits_{r=1}^{N}\mathop{T}\limits_{k=1}^{N}\left[N\left(T\left(\mu_{A_{\mathrm{L}}^k}(\mathbf{v}),\overline{z}_j^k \overset{\neq}{*} \overline{z}_j^r\right)\right)\right]\right]^2}, \tag{8.122}$$

Since t-conorm is nondecreasing and negation N is nonincreasing, we obtain

$$\left.\frac{\partial \underline{\overline{z}}_j}{\partial \mu_{A^l_{\mathrm{L}}}(\mathbf{v})}\right|_{l:\,\overline{z}^l_j=-1} \leq 0, \tag{8.123}$$

In the second step, for (8.114) and $l:\ \overline{z}^l_j = -1$:

$$\left.\frac{\partial \overline{\overline{z}}_j}{\partial \mu_{A^l_{\mathrm{U}}}(\mathbf{v})}\right|_{l:\,\overline{z}^l_j=-1} = \left.\frac{\partial}{\partial \mu_{A^l_{\mathrm{U}}}(\mathbf{v})} \frac{\sum\limits_{r=1}^{N} \overline{z}^r_j \cdot \mathop{T}\limits_{k=1}^{N}\left[N\left(T\left(\mu_{A^k_{\mathrm{U}}}(\mathbf{v}), \overline{z}^k_j \overset{\neq}{*} \overline{z}^r_j\right)\right)\right]}{\sum\limits_{r=1}^{N}\mathop{T}\limits_{k=1}^{N}\left[N\left(T\left(\mu_{A^k_{\mathrm{U}}}(\mathbf{v}), \overline{z}^k_j \overset{\neq}{*} \overline{z}^r_j\right)\right)\right]}\right|_{l:\,\overline{z}^l_j=-1}, \tag{8.124}$$

$$\left.\frac{\partial \overline{\overline{z}}_j}{\partial \mu_{A^l_{\mathrm{U}}}(\mathbf{v})}\right|_{l:\,\overline{z}^l_j=-1} = \frac{\left(\begin{array}{c} \left.\frac{\partial \sum\limits_{r=1}^{N} \overline{z}^r_j \cdot \mathop{T}\limits_{k=1}^{N}\left[N\left(T\left(\mu_{A^k_{\mathrm{U}}}(\mathbf{v}), \overline{z}^k_j \overset{\neq}{*} \overline{z}^r_j\right)\right)\right]}{\partial \mu_{A^l_{\mathrm{U}}}(\mathbf{v})}\right|_{l:\,\overline{z}^l_j=-1} \cdot \\ \sum\limits_{r=1}^{N}\mathop{T}\limits_{k=1}^{N}\left[N\left(T\left(\mu_{A^k_{\mathrm{U}}}(\mathbf{v}), \overline{z}^k_j \overset{\neq}{*} \overline{z}^r_j\right)\right)\right] \\ -\sum\limits_{r=1}^{N} \overline{z}^r_j \cdot \mathop{T}\limits_{k=1}^{N}\left[N\left(T\left(\mu_{A^k_{\mathrm{U}}}(\mathbf{v}), \overline{z}^k_j \overset{\neq}{*} \overline{z}^r_j\right)\right)\right] \cdot \\ \left.\frac{\partial \sum\limits_{r=1}^{N}\mathop{T}\limits_{k=1}^{N}\left[N\left(T\left(\mu_{A^k_{\mathrm{U}}}(\mathbf{v}), \overline{z}^k_j \overset{\neq}{*} \overline{z}^r_j\right)\right)\right]}{\partial \mu_{A^l_{\mathrm{U}}}(\mathbf{v})}\right|_{l:\,\overline{z}^l_j=-1} \end{array}\right)}{\left[\sum\limits_{r=1}^{N}\mathop{T}\limits_{k=1}^{N}\left[N\left(T\left(\mu_{A^k_{\mathrm{U}}}(\mathbf{v}), \overline{z}^k_j \overset{\neq}{*} \overline{z}^r_j\right)\right)\right]\right]^2}, \tag{8.125}$$

$$\left.\frac{\partial \overline{\overline{z}}_j}{\partial \mu_{A^l_{\mathrm{U}}}(\mathbf{v})}\right|_{l:\,\overline{z}^l_j=-1} = \frac{\left(\begin{array}{c} \sum\limits_{r=1}^{N} \overline{z}^r_j \left.\frac{\partial \mathop{T}\limits_{k=1}^{N}\left[N\left(T\left(\mu_{A^k_{\mathrm{U}}}(\mathbf{v}), \overline{z}^k_j \overset{\neq}{*} \overline{z}^r_j\right)\right)\right]}{\partial \mu_{A^l_{\mathrm{U}}}(\mathbf{v})}\right|_{l:\,\overline{z}^l_j=-1} \cdot \\ \sum\limits_{r=1}^{N}\mathop{T}\limits_{k=1}^{N}\left[N\left(T\left(\mu_{A^k_{\mathrm{U}}}(\mathbf{v}), \overline{z}^k_j \overset{\neq}{*} \overline{z}^r_j\right)\right)\right] \\ -\sum\limits_{r=1}^{N} \overline{z}^r_j \cdot \mathop{T}\limits_{k=1}^{N}\left[N\left(T\left(\mu_{A^k_{\mathrm{U}}}(\mathbf{v}), \overline{z}^k_j \overset{\neq}{*} \overline{z}^r_j\right)\right)\right] \cdot \\ \sum\limits_{r=1}^{N} \left.\frac{\partial \mathop{T}\limits_{k=1}^{N}\left[N\left(T\left(\mu_{A^k_{\mathrm{U}}}(\mathbf{v}), \overline{z}^k_j \overset{\neq}{*} \overline{z}^r_j\right)\right)\right]}{\partial \mu_{A^l_{\mathrm{U}}}(\mathbf{v})}\right|_{l:\,\overline{z}^l_j=-1} \end{array}\right)}{\left[\sum\limits_{r=1}^{N}\mathop{T}\limits_{k=1}^{N}\left[N\left(T\left(\mu_{A^k_{\mathrm{U}}}(\mathbf{v}), \overline{z}^k_j \overset{\neq}{*} \overline{z}^r_j\right)\right)\right]\right]^2}, \tag{8.126}$$

On the basis of boundary conditions of t-norms, we get

$$\left.\frac{\partial \overline{\overline{z}}_j}{\partial \mu_{A_{\mathrm{U}}^l}(\mathbf{v})}\right|_{l:\,\overline{z}_j^l=-1} = \frac{\begin{array}{c}\left(\left(\left.\sum\limits_{\substack{r=1\\ r:\,\overline{z}_j^r=1}}^{N} \frac{\partial \underset{\substack{k=1\\ k:\,\overline{z}_j^k=-1}}{\overset{N}{T}} \left[N\left(\mu_{A_{\mathrm{U}}^k}(\mathbf{v})\right)\right]}{\partial \mu_{A_{\mathrm{U}}^l}(\mathbf{v})}\right|_{l:\,\overline{z}_j^l=-1} - \left.\sum\limits_{\substack{r=1\\ r:\,\overline{z}_j^r=-1}}^{N} \frac{\partial \underset{\substack{k=1\\ k:\,\overline{z}_j^k=1}}{\overset{N}{T}} \left[N\left(\mu_{A_{\mathrm{U}}^k}(\mathbf{v})\right)\right]}{\partial \mu_{A_{\mathrm{U}}^l}(\mathbf{v})}\right|_{l:\,\overline{z}_j^l=-1}\right) \cdot \right. \\ \sum\limits_{r=1}^{N} \overset{N}{\underset{k=1}{T}} \left[N\left(T\left(\mu_{A_{\mathrm{U}}^k}(\mathbf{v}), \overline{z}_j^k \overset{\neq}{*} \overline{z}_j^r\right)\right)\right] \\ - \sum\limits_{r=1}^{N} \overline{z}_j^r \cdot \overset{N}{\underset{k=1}{T}} \left[N\left(T\left(\mu_{A_{\mathrm{U}}^k}(\mathbf{v}), \overline{z}_j^k \overset{\neq}{*} \overline{z}_j^r\right)\right)\right] \cdot \\ \left.\left(\left.\sum\limits_{\substack{r=1\\ r:\,\overline{z}_j^r=1}}^{N} \frac{\partial \underset{\substack{k=1\\ k:\,\overline{z}_j^k=-1}}{\overset{N}{T}} \left[N\left(\mu_{A_{\mathrm{U}}^k}(\mathbf{v})\right)\right]}{\partial \mu_{A_{\mathrm{U}}^l}(\mathbf{v})}\right|_{l:\,\overline{z}_j^l=-1} + \left.\sum\limits_{\substack{r=1\\ r:\,\overline{z}_j^r=-1}}^{N} \frac{\partial \underset{\substack{k=1\\ k:\,\overline{z}_j^k=1}}{\overset{N}{T}} \left[N\left(\mu_{A_{\mathrm{U}}^k}(\mathbf{v})\right)\right]}{\partial \mu_{A_{\mathrm{U}}^l}(\mathbf{v})}\right|_{l:\,\overline{z}_j^l=-1}\right)\right)\end{array}}{\left[\sum\limits_{r=1}^{N} \overset{N}{\underset{k=1}{T}} \left[N\left(T\left(\mu_{A_{\mathrm{U}}^k}(\mathbf{v}), \overline{z}_j^k \overset{\neq}{*} \overline{z}_j^r\right)\right)\right]\right]^2}, \tag{8.127}$$

$$\left.\frac{\partial \overline{\overline{z}}_j}{\partial \mu_{A_{\mathrm{U}}^l}(\mathbf{v})}\right|_{l:\,\overline{z}_j^l=-1} = \frac{\begin{array}{c}\left(\left(\left.\sum\limits_{\substack{r=1\\ r:\,\overline{z}_j^r=1}}^{N} \frac{\partial \underset{\substack{k=1\\ k:\,\overline{z}_j^k=-1}}{\overset{N}{T}} \left[N\left(\mu_{A_{\mathrm{U}}^k}(\mathbf{v})\right)\right]}{\partial \mu_{A_{\mathrm{U}}^l}(\mathbf{v})}\right|_{l:\,\overline{z}_j^l=-1} - 0\right) \cdot \right. \\ \sum\limits_{r=1}^{N} \overset{N}{\underset{k=1}{T}} \left[N\left(T\left(\mu_{A_{\mathrm{U}}^k}(\mathbf{v}), \overline{z}_j^k \overset{\neq}{*} \overline{z}_j^r\right)\right)\right] \\ - \sum\limits_{r=1}^{N} \overline{z}_j^r \cdot \overset{N}{\underset{k=1}{T}} \left[N\left(T\left(\mu_{A_{\mathrm{U}}^k}(\mathbf{v}), \overline{z}_j^k \overset{\neq}{*} \overline{z}_j^r\right)\right)\right] \cdot \\ \left.\left(\left.\sum\limits_{\substack{r=1\\ r:\,\overline{z}_j^r=1}}^{N} \frac{\partial \underset{\substack{k=1\\ k:\,\overline{z}_j^k=-1}}{\overset{N}{T}} \left[N\left(\mu_{A_{\mathrm{U}}^k}(\mathbf{v})\right)\right]}{\partial \mu_{A_{\mathrm{U}}^l}(\mathbf{v})}\right|_{l:\,\overline{z}_j^l=-1} + 0\right)\right)\end{array}}{\left[\sum\limits_{r=1}^{N} \overset{N}{\underset{k=1}{T}} \left[N\left(T\left(\mu_{A_{\mathrm{U}}^k}(\mathbf{v}), \overline{z}_j^k \overset{\neq}{*} \overline{z}_j^r\right)\right)\right]\right]^2}, \tag{8.128}$$

$$\left.\frac{\partial \overline{\overline{z}}_j}{\partial \mu_{A_{\mathrm{U}}^l}(\mathbf{v})}\right|_{l:\,\overline{z}_j^l=-1} = \frac{\begin{array}{c}\left(\left.\sum\limits_{\substack{r=1\\ r:\,\overline{z}_j^r=1}}^{N} \frac{\partial \underset{\substack{k=1\\ k:\,\overline{z}_j^k=-1}}{\overset{N}{T}} \left[N\left(\mu_{A_{\mathrm{U}}^k}(\mathbf{v})\right)\right]}{\partial \mu_{A_{\mathrm{U}}^l}(\mathbf{v})}\right|_{l:\,\overline{z}_j^l=-1} \cdot \right. \\ \left. \cdot \sum\limits_{r=1}^{N} \left(1 - \overline{z}_j^r\right) \cdot \overset{N}{\underset{k=1}{T}} \left[N\left(T\left(\mu_{A_{\mathrm{U}}^k}(\mathbf{v}), \overline{z}_j^k \overset{\neq}{*} \overline{z}_j^r\right)\right)\right]\right)\end{array}}{\left[\sum\limits_{r=1}^{N} \overset{N}{\underset{k=1}{T}} \left[N\left(T\left(\mu_{A_{\mathrm{U}}^k}(\mathbf{v}), \overline{z}_j^k \overset{\neq}{*} \overline{z}_j^r\right)\right)\right]\right]^2}, \tag{8.129}$$

Since t-conorm is nondecreasing and negation N is nonincreasing we obtain

$$\left.\frac{\partial \overline{\overline{z}_j}}{\partial \mu_{A_{\mathrm{U}}^l}(\mathbf{v})}\right|_{l:\, \overline{z}_j^l=-1} \leq 0, \tag{8.130}$$

In the third step for (8.113) and $l\colon \overline{z}_j^l = 1$:

$$\left.\frac{\partial \underline{\overline{z}_j}}{\partial \mu_{A_{\mathrm{L}}^l}(\mathbf{v})}\right|_{l:\, \overline{z}_j^l=1} = \left.\frac{\partial}{\partial \mu_{A_{\mathrm{L}}^l}(\mathbf{v})} \frac{\sum\limits_{r=1}^{N} \overline{z}_j^r \cdot \mathop{T}\limits_{k=1}^{N} \left[N\left(T\left(\mu_{A_{\mathrm{L}}^k}(\mathbf{v}), \overline{z}_j^k \overset{\neq}{*} \overline{z}_j^r\right)\right)\right]}{\sum\limits_{r=1}^{N} \mathop{T}\limits_{k=1}^{N} \left[N\left(T\left(\mu_{A_{\mathrm{L}}^k}(\mathbf{v}), \overline{z}_j^k \overset{\neq}{*} \overline{z}_j^r\right)\right)\right]}\right|_{l:\, \overline{z}_j^l=1}, \tag{8.131}$$

$$\left.\frac{\partial \underline{\overline{z}_j}}{\partial \mu_{A_{\mathrm{L}}^l}(\mathbf{v})}\right|_{l:\, \overline{z}_j^l=1} = \frac{\left(\begin{array}{c} \left.\frac{\partial \sum\limits_{r=1}^{N} \overline{z}_j^r \cdot \mathop{T}\limits_{k=1}^{N} \left[N\left(T\left(\mu_{A_{\mathrm{L}}^k}(\mathbf{v}), \overline{z}_j^k \overset{\neq}{*} \overline{z}_j^r\right)\right)\right]}{\partial \mu_{A_{\mathrm{L}}^l}(\mathbf{v})}\right|_{l:\, \overline{z}_j^l=1} \cdot \\ \sum\limits_{r=1}^{N} \mathop{T}\limits_{k=1}^{N} \left[N\left(T\left(\mu_{A_{\mathrm{L}}^k}(\mathbf{v}), \overline{z}_j^k \overset{\neq}{*} \overline{z}_j^r\right)\right)\right] \\ - \sum\limits_{r=1}^{N} \overline{z}_j^r \cdot \mathop{T}\limits_{k=1}^{N} \left[N\left(T\left(\mu_{A_{\mathrm{L}}^k}(\mathbf{v}), \overline{z}_j^k \overset{\neq}{*} \overline{z}_j^r\right)\right)\right] \cdot \\ \left.\frac{\partial \sum\limits_{r=1}^{N} \mathop{T}\limits_{k=1}^{N} \left[N\left(T\left(\mu_{A_{\mathrm{L}}^k}(\mathbf{v}), \overline{z}_j^k \overset{\neq}{*} \overline{z}_j^r\right)\right)\right]}{\partial \mu_{A_{\mathrm{L}}^l}(\mathbf{v})}\right|_{l:\, \overline{z}_j^l=1} \end{array}\right)}{\left[\sum\limits_{r=1}^{N} \mathop{T}\limits_{k=1}^{N} \left[N\left(T\left(\mu_{A_{\mathrm{L}}^k}(\mathbf{v}), \overline{z}_j^k \overset{\neq}{*} \overline{z}_j^r\right)\right)\right]\right]^2}, \tag{8.132}$$

$$\left.\frac{\partial \underline{\overline{z}_j}}{\partial \mu_{A_{\mathrm{L}}^l}(\mathbf{v})}\right|_{l:\, \overline{z}_j^l=1} = \frac{\left(\begin{array}{c} \sum\limits_{r=1}^{N} \overline{z}_j^r \left.\frac{\partial \mathop{T}\limits_{k=1}^{N} \left[N\left(T\left(\mu_{A_{\mathrm{L}}^k}(\mathbf{v}), \overline{z}_j^k \overset{\neq}{*} \overline{z}_j^r\right)\right)\right]}{\partial \mu_{A_{\mathrm{L}}^l}(\mathbf{v})}\right|_{l:\, \overline{z}_j^l=1} \cdot \\ \sum\limits_{r=1}^{N} \mathop{T}\limits_{k=1}^{N} \left[N\left(T\left(\mu_{A_{\mathrm{L}}^k}(\mathbf{v}), \overline{z}_j^k \overset{\neq}{*} \overline{z}_j^r\right)\right)\right] \\ - \sum\limits_{r=1}^{N} \overline{z}_j^r \cdot \mathop{T}\limits_{k=1}^{N} \left[N\left(T\left(\mu_{A_{\mathrm{L}}^k}(\mathbf{v}), \overline{z}_j^k \overset{\neq}{*} \overline{z}_j^r\right)\right)\right] \cdot \\ \sum\limits_{r=1}^{N} \left.\frac{\partial \mathop{T}\limits_{k=1}^{N} \left[N\left(T\left(\mu_{A_{\mathrm{L}}^k}(\mathbf{v}), \overline{z}_j^k \overset{\neq}{*} \overline{z}_j^r\right)\right)\right]}{\partial \mu_{A_{\mathrm{L}}^l}(\mathbf{v})}\right|_{l:\, \overline{z}_j^l=1} \end{array}\right)}{\left[\sum\limits_{r=1}^{N} \mathop{T}\limits_{k=1}^{N} \left[N\left(T\left(\mu_{A_{\mathrm{L}}^k}(\mathbf{v}), \overline{z}_j^k \overset{\neq}{*} \overline{z}_j^r\right)\right)\right]\right]^2}, \tag{8.133}$$

On the basis of boundary conditions of t-norms, we get

$$\left.\frac{\partial \underline{z}_j}{\partial \mu_{A_{\mathrm{L}}^l}(\mathbf{v})}\right|_{l:\,\overline{z}_j^l=1} = \frac{\left(\begin{array}{c}\left(\sum\limits_{\substack{r=1\\ r:\,\overline{z}_j^r=1}}^{N} \left.\frac{\partial \underset{\substack{k=1\\ k:\,\overline{z}_j^k=-1}}{\overset{N}{T}} \left[N\left(\mu_{A_{\mathrm{L}}^k}(\mathbf{v})\right)\right]}{\partial \mu_{A_{\mathrm{L}}^l}(\mathbf{v})}\right|_{l:\,\overline{z}_j^l=1} - \sum\limits_{\substack{r=1\\ r:\,\overline{z}_j^r=-1}}^{N} \left.\frac{\partial \underset{\substack{k=1\\ k:\,\overline{z}_j^k=1}}{\overset{N}{T}} \left[N\left(\mu_{A_{\mathrm{L}}^k}(\mathbf{v})\right)\right]}{\partial \mu_{A_{\mathrm{L}}^l}(\mathbf{v})}\right|_{l:\,\overline{z}_j^l=1}\right) \cdot \\ \sum\limits_{r=1}^{N} \overset{N}{\underset{k=1}{T}} \left[N\left(T\left(\mu_{A_{\mathrm{L}}^k}(\mathbf{v}), \overline{z}_j^k \overset{\neq}{*} \overline{z}_j^r\right)\right)\right] \\ - \sum\limits_{r=1}^{N} \overline{z}_j^r \cdot \overset{N}{\underset{k=1}{T}} \left[N\left(T\left(\mu_{A_{\mathrm{L}}^k}(\mathbf{v}), \overline{z}_j^k \overset{\neq}{*} \overline{z}_j^r\right)\right)\right] \cdot \\ \left(\sum\limits_{\substack{r=1\\ r:\,\overline{z}_j^r=1}}^{N} \left.\frac{\partial \underset{\substack{k=1\\ k:\,\overline{z}_j^k=-1}}{\overset{N}{T}} \left[N\left(\mu_{A_{\mathrm{L}}^k}(\mathbf{v})\right)\right]}{\partial \mu_{A_{\mathrm{L}}^l}(\mathbf{v})}\right|_{l:\,\overline{z}_j^l=1} + \sum\limits_{\substack{r=1\\ r:\,\overline{z}_j^r=-1}}^{N} \left.\frac{\partial \underset{\substack{k=1\\ k:\,\overline{z}_j^k=1}}{\overset{N}{T}} \left[N\left(\mu_{A_{\mathrm{L}}^k}(\mathbf{v})\right)\right]}{\partial \mu_{A_{\mathrm{L}}^l}(\mathbf{v})}\right|_{l:\,\overline{z}_j^l=1}\right)\end{array}\right)}{\left[\sum\limits_{r=1}^{N} \overset{N}{\underset{k=1}{T}} \left[N\left(T\left(\mu_{A_{\mathrm{L}}^k}(\mathbf{v}), \overline{z}_j^k \overset{\neq}{*} \overline{z}_j^r\right)\right)\right]\right]^2}, \tag{8.134}$$

$$\left.\frac{\partial \underline{z}_j}{\partial \mu_{A_{\mathrm{L}}^l}(\mathbf{v})}\right|_{l:\,\overline{z}_j^l=1} = \frac{\left(\begin{array}{c}\left(0 - \sum\limits_{\substack{r=1\\ r:\,\overline{z}_j^r=-1}}^{N} \left.\frac{\partial \underset{\substack{k=1\\ k:\,\overline{z}_j^k=1}}{\overset{N}{T}} \left[N\left(\mu_{A_{\mathrm{L}}^k}(\mathbf{v})\right)\right]}{\partial \mu_{A_{\mathrm{L}}^l}(\mathbf{v})}\right|_{l:\,\overline{z}_j^l=1}\right) \cdot \\ \sum\limits_{r=1}^{N} \overset{N}{\underset{k=1}{T}} \left[N\left(T\left(\mu_{A_{\mathrm{L}}^k}(\mathbf{v}), \overline{z}_j^k \overset{\neq}{*} \overline{z}_j^r\right)\right)\right] \\ - \sum\limits_{r=1}^{N} \overline{z}_j^r \cdot \overset{N}{\underset{k=1}{T}} \left[N\left(T\left(\mu_{A_{\mathrm{L}}^k}(\mathbf{v}), \overline{z}_j^k \overset{\neq}{*} \overline{z}_j^r\right)\right)\right] \cdot \\ \left(0 + \sum\limits_{\substack{r=1\\ r:\,\overline{z}_j^r=-1}}^{N} \left.\frac{\partial \underset{\substack{k=1\\ k:\,\overline{z}_j^k=1}}{\overset{N}{T}} \left[N\left(\mu_{A_{\mathrm{L}}^k}(\mathbf{v})\right)\right]}{\partial \mu_{A_{\mathrm{L}}^l}(\mathbf{v})}\right|_{l:\,\overline{z}_j^l=1}\right)\end{array}\right)}{\left[\sum\limits_{r=1}^{N} \overset{N}{\underset{k=1}{T}} \left[N\left(T\left(\mu_{A_{\mathrm{L}}^k}(\mathbf{v}), \overline{z}_j^k \overset{\neq}{*} \overline{z}_j^r\right)\right)\right]\right]^2}, \tag{8.135}$$

$$\left.\frac{\partial \underline{z}_j}{\partial \mu_{A_{\mathrm{L}}^l}(\mathbf{v})}\right|_{l:\,\overline{z}_j^l=1} = -\frac{\left(\begin{array}{c}\sum\limits_{\substack{r=1\\ r:\,\overline{z}_j^r=-1}}^{N} \left.\frac{\partial \underset{\substack{k=1\\ k:\,\overline{z}_j^k=1}}{\overset{N}{T}} \left[N\left(\mu_{A_{\mathrm{L}}^k}(\mathbf{v})\right)\right]}{\partial \mu_{A_{\mathrm{L}}^l}(\mathbf{v})}\right|_{l:\,\overline{z}_j^l=1} \cdot \\ \sum\limits_{r=1}^{N} \left(1 + \overline{z}_j^r\right) \cdot \overset{N}{\underset{k=1}{T}} \left[N\left(T\left(\mu_{A_{\mathrm{L}}^k}(\mathbf{v}), \overline{z}_j^k \overset{\neq}{*} \overline{z}_j^r\right)\right)\right]\end{array}\right)}{\left[\sum\limits_{r=1}^{N} \overset{N}{\underset{k=1}{T}} \left[N\left(T\left(\mu_{A_{\mathrm{L}}^k}(\mathbf{v}), \overline{z}_j^k \overset{\neq}{*} \overline{z}_j^r\right)\right)\right]\right]^2}, \tag{8.136}$$

Since T–norm is nondecreasing and negation N is nonincreasing, we obtain

$$\left.\frac{\partial \underline{z}_j}{\partial \mu_{A_{\mathrm{L}}^l}(\mathbf{v})}\right|_{l:\, \overline{z}_j^l=1} \geq 0, \tag{8.137}$$

In the fourth step for (8.114) and $l\colon \overline{z}_j^l = 1$:

$$\left.\frac{\partial \overline{\overline{z}}_j}{\partial \mu_{A_{\mathrm{U}}^l}(\mathbf{v})}\right|_{l:\, \overline{z}_j^l=1} = \frac{\partial}{\partial \mu_{A_{\mathrm{U}}^l}(\mathbf{v})} \left.\frac{\sum\limits_{r=1}^{N} \overline{z}_j^r \cdot \mathop{T}\limits_{k=1}^{N} \left[N\left(T\left(\mu_{A_{\mathrm{U}}^k}(\mathbf{v}), \overline{z}_j^k \overset{\neq}{*} \overline{z}_j^r\right)\right)\right]}{\sum\limits_{r=1}^{N} \mathop{T}\limits_{k=1}^{N} \left[N\left(T\left(\mu_{A_{\mathrm{U}}^k}(\mathbf{v}), \overline{z}_j^k \overset{\neq}{*} \overline{z}_j^r\right)\right)\right]}\right|_{l:\, \overline{z}_j^l=1}, \tag{8.138}$$

$$\left.\frac{\partial \overline{\overline{z}}_j}{\partial \mu_{A_{\mathrm{U}}^l}(\mathbf{v})}\right|_{l:\, \overline{z}_j^l=1} = \frac{\left(\begin{array}{c} \left.\dfrac{\partial \sum\limits_{r=1}^{N} \overline{z}_j^r \cdot \mathop{T}\limits_{k=1}^{N} \left[N\left(T\left(\mu_{A_{\mathrm{U}}^k}(\mathbf{v}), \overline{z}_j^k \overset{\neq}{*} \overline{z}_j^r\right)\right)\right]}{\partial \mu_{A_{\mathrm{U}}^l}(\mathbf{v})}\right|_{l:\, \overline{z}_j^l=1} \cdot \\ \sum\limits_{r=1}^{N} \mathop{T}\limits_{k=1}^{N} \left[N\left(T\left(\mu_{A_{\mathrm{U}}^k}(\mathbf{v}), \overline{z}_j^k \overset{\neq}{*} \overline{z}_j^r\right)\right)\right] \\ - \sum\limits_{r=1}^{N} \overline{z}_j^r \cdot \mathop{T}\limits_{k=1}^{N} \left[N\left(T\left(\mu_{A_{\mathrm{U}}^k}(\mathbf{v}), \overline{z}_j^k \overset{\neq}{*} \overline{z}_j^r\right)\right)\right] \cdot \\ \left.\dfrac{\partial \sum\limits_{r=1}^{N} \mathop{T}\limits_{k=1}^{N} \left[N\left(T\left(\mu_{A_{\mathrm{U}}^k}(\mathbf{v}), \overline{z}_j^k \overset{\neq}{*} \overline{z}_j^r\right)\right)\right]}{\partial \mu_{A_{\mathrm{U}}^l}(\mathbf{v})}\right|_{l:\, \overline{z}_j^l=1} \end{array}\right)}{\left[\sum\limits_{r=1}^{N} \mathop{T}\limits_{k=1}^{N} \left[N\left(T\left(\mu_{A_{\mathrm{U}}^k}(\mathbf{v}), \overline{z}_j^k \overset{\neq}{*} \overline{z}_j^r\right)\right)\right]\right]^2}, \tag{8.139}$$

$$\left.\frac{\partial \overline{\overline{z}}_j}{\partial \mu_{A_{\mathrm{U}}^l}(\mathbf{v})}\right|_{l:\, \overline{z}_j^l=1} = \frac{\left(\begin{array}{c} \sum\limits_{r=1}^{N} \overline{z}_j^r \left.\dfrac{\partial \mathop{T}\limits_{k=1}^{N} \left[N\left(T\left(\mu_{A_{\mathrm{U}}^k}(\mathbf{v}), \overline{z}_j^k \overset{\neq}{*} \overline{z}_j^r\right)\right)\right]}{\partial \mu_{A_{\mathrm{U}}^l}(\mathbf{v})}\right|_{l:\, \overline{z}_j^l=1} \cdot \\ \sum\limits_{r=1}^{N} \mathop{T}\limits_{k=1}^{N} \left[N\left(T\left(\mu_{A_{\mathrm{U}}^k}(\mathbf{v}), \overline{z}_j^k \overset{\neq}{*} \overline{z}_j^r\right)\right)\right] \\ - \sum\limits_{r=1}^{N} \overline{z}_j^r \cdot \mathop{T}\limits_{k=1}^{N} \left[N\left(T\left(\mu_{A_{\mathrm{U}}^k}(\mathbf{v}), \overline{z}_j^k \overset{\neq}{*} \overline{z}_j^r\right)\right)\right] \cdot \\ \sum\limits_{r=1}^{N} \left.\dfrac{\partial \mathop{T}\limits_{k=1}^{N} \left[N\left(T\left(\mu_{A_{\mathrm{U}}^k}(\mathbf{v}), \overline{z}_j^k \overset{\neq}{*} \overline{z}_j^r\right)\right)\right]}{\partial \mu_{A_{\mathrm{U}}^l}(\mathbf{v})}\right|_{l:\, \overline{z}_j^l=1} \end{array}\right)}{\left[\sum\limits_{r=1}^{N} \mathop{T}\limits_{k=1}^{N} \left[N\left(T\left(\mu_{A_{\mathrm{U}}^k}(\mathbf{v}), \overline{z}_j^k \overset{\neq}{*} \overline{z}_j^r\right)\right)\right]\right]^2}, \tag{8.140}$$

On the basis of boundary conditions of t-norms, we get

$$\left.\frac{\partial\overline{\overline{z}}_j}{\partial\mu_{A_U^l}(\mathbf{v})}\right|_{l:\,\overline{z}_j^l=1}=\frac{\left(\begin{array}{c}\left(\sum\limits_{\substack{r=1\\ r:\,\overline{z}_j^r=1}}^{N}\left.\frac{\partial\mathop{T}\limits_{\substack{k=1\\ k:\,\overline{z}_j^k=-1}}^{N}\left[N\left(\mu_{A_U^k}(\mathbf{v})\right)\right]}{\partial\mu_{A_U^l}(\mathbf{v})}\right|_{l:\,\overline{z}_j^l=1}-\sum\limits_{\substack{r=1\\ r:\,\overline{z}_j^r=-1}}^{N}\left.\frac{\partial\mathop{T}\limits_{\substack{k=1\\ k:\,\overline{z}_j^k=1}}^{N}\left[N\left(\mu_{A_U^k}(\mathbf{v})\right)\right]}{\partial\mu_{A_U^l}(\mathbf{v})}\right|_{l:\,\overline{z}_j^l=1}\right)\cdot\\ \sum\limits_{r=1}^{N}\mathop{T}\limits_{k=1}^{N}\left[N\left(T\left(\mu_{A_U^k}(\mathbf{v}),\overline{z}_j^k\overset{\neq}{*}\overline{z}_j^r\right)\right)\right]\\ -\sum\limits_{r=1}^{N}\overline{z}_j^r\cdot\mathop{T}\limits_{k=1}^{N}\left[N\left(T\left(\mu_{A_U^k}(\mathbf{v}),\overline{z}_j^k\overset{\neq}{*}\overline{z}_j^r\right)\right)\right]\cdot\\ \left(\sum\limits_{\substack{r=1\\ r:\,\overline{z}_j^r=1}}^{N}\left.\frac{\partial\mathop{T}\limits_{\substack{k=1\\ k:\,\overline{z}_j^k=-1}}^{N}\left[N\left(\mu_{A_U^k}(\mathbf{v})\right)\right]}{\partial\mu_{A_U^l}(\mathbf{v})}\right|_{l:\,\overline{z}_j^l=1}+\sum\limits_{\substack{r=1\\ r:\,\overline{z}_j^r=-1}}^{N}\left.\frac{\partial\mathop{T}\limits_{\substack{k=1\\ k:\,\overline{z}_j^k=1}}^{N}\left[N\left(\mu_{A_U^k}(\mathbf{v})\right)\right]}{\partial\mu_{A_U^l}(\mathbf{v})}\right|_{l:\,\overline{z}_j^l=1}\right)\end{array}\right)}{\left[\sum\limits_{r=1}^{N}\mathop{T}\limits_{k=1}^{N}\left[N\left(T\left(\mu_{A_U^k}(\mathbf{v}),\overline{z}_j^k\overset{\neq}{*}\overline{z}_j^r\right)\right)\right]\right]^2},\tag{8.141}$$

$$\left.\frac{\partial\overline{\overline{z}}_j}{\partial\mu_{A_U^l}(\mathbf{v})}\right|_{l:\,\overline{z}_j^l=1}=\frac{\left(\begin{array}{c}\left(0-\sum\limits_{\substack{r=1\\ r:\,\overline{z}_j^r=-1}}^{N}\left.\frac{\partial\mathop{T}\limits_{\substack{k=1\\ k:\,\overline{z}_j^k=1}}^{N}\left[N\left(\mu_{A_U^k}(\mathbf{v})\right)\right]}{\partial\mu_{A_U^l}(\mathbf{v})}\right|_{l:\,\overline{z}_j^l=1}\right)\cdot\\ \sum\limits_{r=1}^{N}\mathop{T}\limits_{k=1}^{N}\left[N\left(T\left(\mu_{A_U^k}(\mathbf{v}),\overline{z}_j^k\overset{\neq}{*}\overline{z}_j^r\right)\right)\right]\\ -\sum\limits_{r=1}^{N}\overline{z}_j^r\cdot\mathop{T}\limits_{k=1}^{N}\left[N\left(T\left(\mu_{A_U^k}(\mathbf{v}),\overline{z}_j^k\overset{\neq}{*}\overline{z}_j^r\right)\right)\right]\cdot\\ \left(0+\sum\limits_{\substack{r=1\\ r:\,\overline{z}_j^r=-1}}^{N}\left.\frac{\partial\mathop{T}\limits_{\substack{k=1\\ k:\,\overline{z}_j^k=1}}^{N}\left[N\left(\mu_{A_U^k}(\mathbf{v})\right)\right]}{\partial\mu_{A_U^l}(\mathbf{v})}\right|_{l:\,\overline{z}_j^l=1}\right)\end{array}\right)}{\left[\sum\limits_{r=1}^{N}\mathop{T}\limits_{k=1}^{N}\left[N\left(T\left(\mu_{A_U^k}(\mathbf{v}),\overline{z}_j^k\overset{\neq}{*}\overline{z}_j^r\right)\right)\right]\right]^2},\tag{8.142}$$

$$\left.\frac{\partial\overline{\overline{z}}_j}{\partial\mu_{A_U^l}(\mathbf{v})}\right|_{l:\,\overline{z}_j^l=1}=-\frac{\left(\begin{array}{c}\sum\limits_{\substack{r=1\\ r:\,\overline{z}_j^r=-1}}^{N}\left.\frac{\partial\mathop{T}\limits_{\substack{k=1\\ k:\,\overline{z}_j^k=1}}^{N}\left[N\left(\mu_{A_U^k}(\mathbf{v})\right)\right]}{\partial\mu_{A_U^l}(\mathbf{v})}\right|_{l:\,\overline{z}_j^l=1}\cdot\\ \sum\limits_{r=1}^{N}\left(1+\overline{z}_j^r\right)\cdot\mathop{T}\limits_{k=1}^{N}\left[N\left(T\left(\mu_{A_U^k}(\mathbf{v}),\overline{z}_j^k\overset{\neq}{*}\overline{z}_j^r\right)\right)\right]\end{array}\right)}{\left[\sum\limits_{r=1}^{N}\mathop{T}\limits_{k=1}^{N}\left[N\left(T\left(\mu_{A_U^k}(\mathbf{v}),\overline{z}_j^k\overset{\neq}{*}\overline{z}_j^r\right)\right)\right]\right]^2},\tag{8.143}$$

Since t-norm is nondecreasing, and negation N is nonincreasing we obtain

$$\left.\frac{\partial \overline{\overline{z}}_j}{\partial \mu_{A_{\mathrm{U}}^l}(\mathbf{v})}\right|_{l:\,\overline{z}_j^l=1} \geq 0, \tag{8.144}$$

Analyzing inequalities (8.123), (8.130), (8.137) and (8.144) it is easily seen that in formulas (8.113) and (8.114) fuzzy sets A_{L}^r and A_{U}^r should be chosen according to descriptions (8.115) and (8.116).

8.3.4 R–DCOG Rough Fuzzy Classifier

Following formula (8.47), the lower and upper approximations of the membership of object x to class ω_j are given by

$$\underline{\overline{z}}_j = \frac{\sum\limits_{r=1}^{N} \overline{z}_j^r \cdot \mathop{T}\limits_{k=1}^{N} \left[\left(\mu_{A_{\mathrm{L}}^k}(\mathbf{v}) \overset{=}{*} 0 \right) \cdot \left(\overline{z}_j^k \overset{\neq}{*} \overline{z}_j^r \right) + \left(\overline{z}_j^k \overset{=}{*} \overline{z}_j^r \right) \right]}{\sum\limits_{r=1}^{N} \mathop{T}\limits_{k=1}^{N} \left[\left(\mu_{A_{\mathrm{L}}^k}(\mathbf{v}) \overset{=}{*} 0 \right) \cdot \left(\overline{z}_j^k \overset{\neq}{*} \overline{z}_j^r \right) + \left(\overline{z}_j^k \overset{=}{*} \overline{z}_j^r \right) \right]}, \tag{8.145}$$

$$\overline{\overline{z}}_j = \frac{\sum\limits_{r=1}^{N} \overline{z}_j^r \cdot \mathop{T}\limits_{k=1}^{N} \left[\left(\mu_{A_{\mathrm{U}}^k}(\mathbf{v}) \overset{=}{*} 0 \right) \cdot \left(\overline{z}_j^k \overset{\neq}{*} \overline{z}_j^r \right) + \left(\overline{z}_j^k \overset{=}{*} \overline{z}_j^r \right) \right]}{\sum\limits_{r=1}^{N} \mathop{T}\limits_{k=1}^{N} \left[\left(\mu_{A_{\mathrm{U}}^k}(\mathbf{v}) \overset{=}{*} 0 \right) \cdot \left(\overline{z}_j^k \overset{\neq}{*} \overline{z}_j^r \right) + \left(\overline{z}_j^k \overset{=}{*} \overline{z}_j^r \right) \right]}. \tag{8.146}$$

Theorem 8.4. *Fuzzy sets A_{L}^r and A_{U}^r in descriptions* (8.145) *and* (8.146) *should be chosen as follows*

$$A_{\mathrm{L}}^r = \begin{cases} \underline{\widetilde{P}A^r} & \text{if } \overline{\mathrm{z}}_{\mathrm{j}}^{\mathrm{r}} = 1 \\ \overline{\widetilde{P}A^r} & \text{if } \overline{\mathrm{z}}_{\mathrm{j}}^{\mathrm{r}} = -1. \end{cases} \tag{8.147}$$

$$A_{\mathrm{U}}^r = \begin{cases} \overline{\widetilde{P}A^r} & \text{if } \overline{\mathrm{z}}_{\mathrm{j}}^{\mathrm{r}} = 1 \\ \underline{\widetilde{P}A^r} & \text{if } \overline{\mathrm{z}}_{\mathrm{j}}^{\mathrm{r}} = -1. \end{cases} \tag{8.148}$$

Proof (Theorem 8.4).
In the first step, for (8.147) and $l\colon \overline{z}_j^l = -1$:

$$\left.\frac{\partial \underline{\overline{z}}_j}{\partial \mu_{A_{\mathrm{L}}^l}(\mathbf{v})}\right|_{l:\,\overline{z}_j^l=-1} = \frac{\partial}{\partial \mu_{A_{\mathrm{L}}^l}(\mathbf{v})} \left.\frac{\sum\limits_{r=1}^{N} \overline{z}_j^r \cdot \mathop{T}\limits_{k=1}^{N} \left[\left(\mu_{A_{\mathrm{L}}^k}(\mathbf{v}) \overset{=}{*} 0 \right) \cdot \left(\overline{z}_j^k \overset{\neq}{*} \overline{z}_j^r \right) + \left(\overline{z}_j^k \overset{=}{*} \overline{z}_j^r \right) \right]}{\sum\limits_{r=1}^{N} \mathop{T}\limits_{k=1}^{N} \left[\left(\mu_{A_{\mathrm{L}}^k}(\mathbf{v}) \overset{=}{*} 0 \right) \cdot \left(\overline{z}_j^k \overset{\neq}{*} \overline{z}_j^r \right) + \left(\overline{z}_j^k \overset{=}{*} \overline{z}_j^r \right) \right]}\right|_{l:\,\overline{z}_j^l=-1}, \tag{8.149}$$

$$
\left.\frac{\partial \underline{\overline{z}}_j}{\partial \mu_{A_{\mathrm{L}}^l}(\mathbf{v})}\right|_{l:\,\overline{z}_j^l=-1} = \frac{\left(\begin{array}{l}
\left.\dfrac{\partial \sum\limits_{r=1}^{N} \overline{z}_j^r \cdot \mathop{T}\limits_{k=1}^{N}\left[\left(\mu_{A_{\mathrm{L}}^k}(\mathbf{v}) \overset{=}{*} 0\right)\cdot\left(\overline{z}_j^k \overset{\neq}{*} \overline{z}_j^r\right)+\left(\overline{z}_j^k \overset{=}{*} \overline{z}_j^r\right)\right]}{\partial \mu_{A_{\mathrm{L}}^l}(\mathbf{v})}\right|_{l:\,\overline{z}_j^l=-1} \cdot \\
\cdot \sum\limits_{r=1}^{N} \mathop{T}\limits_{k=1}^{N}\left[\left(\mu_{A_{\mathrm{L}}^k}(\mathbf{v}) \overset{=}{*} 0\right)\cdot\left(\overline{z}_j^k \overset{\neq}{*} \overline{z}_j^r\right)+\left(\overline{z}_j^k \overset{=}{*} \overline{z}_j^r\right)\right] \\
-\sum\limits_{r=1}^{N} \overline{z}_j^r \cdot \mathop{T}\limits_{k=1}^{N}\left[\left(\mu_{A_{\mathrm{L}}^k}(\mathbf{v}) \overset{=}{*} 0\right)\cdot\left(\overline{z}_j^k \overset{\neq}{*} \overline{z}_j^r\right)+\left(\overline{z}_j^k \overset{=}{*} \overline{z}_j^r\right)\right] \\
\left.\dfrac{\partial \sum\limits_{r=1}^{N} \mathop{T}\limits_{k=1}^{N}\left[\left(\mu_{A_{\mathrm{L}}^k}(\mathbf{v}) \overset{=}{*} 0\right)\cdot\left(\overline{z}_j^k \overset{\neq}{*} \overline{z}_j^r\right)+\left(\overline{z}_j^k \overset{=}{*} \overline{z}_j^r\right)\right]}{\partial \mu_{A_{\mathrm{L}}^l}(\mathbf{v})}\right|_{l:\,\overline{z}_j^l=-1}
\end{array}\right)}{\left[\sum\limits_{r=1}^{N} \mathop{T}\limits_{k=1}^{N}\left[\left(\mu_{A_{\mathrm{L}}^k}(\mathbf{v}) \overset{=}{*} 0\right)\cdot\left(\overline{z}_j^k \overset{\neq}{*} \overline{z}_j^r\right)+\left(\overline{z}_j^k \overset{=}{*} \overline{z}_j^r\right)\right]\right]^2}, \tag{8.150}
$$

$$
\left.\frac{\partial \underline{\overline{z}}_j}{\partial \mu_{A_{\mathrm{L}}^l}(\mathbf{v})}\right|_{l:\,\overline{z}_j^l=-1} = \frac{\left(\begin{array}{l}
\left(\begin{array}{l}
\sum\limits_{\substack{r=1 \\ r:\,\overline{z}_j^r=1}}^{N} \left.\dfrac{\partial \mathop{T}\limits_{k=1}^{N}\left[\left(\mu_{A_{\mathrm{L}}^k}(\mathbf{v}) \overset{=}{*} 0\right)\cdot\left(\overline{z}_j^k \overset{\neq}{*} \overline{z}_j^r\right)+\left(\overline{z}_j^k \overset{=}{*} \overline{z}_j^r\right)\right]}{\partial \mu_{A_{\mathrm{L}}^l}(\mathbf{v})}\right|_{l:\,\overline{z}_j^l=-1} \\
-\sum\limits_{\substack{r=1 \\ r:\,\overline{z}_j^r=-1}}^{N} \left.\dfrac{\partial \mathop{T}\limits_{k=1}^{N}\left[\left(\mu_{A_{\mathrm{L}}^k}(\mathbf{v}) \overset{=}{*} 0\right)\cdot\left(\overline{z}_j^k \overset{\neq}{*} \overline{z}_j^r\right)+\left(\overline{z}_j^k \overset{=}{*} \overline{z}_j^r\right)\right]}{\partial \mu_{A_{\mathrm{L}}^l}(\mathbf{v})}\right|_{l:\,\overline{z}_j^l=-1}
\end{array}\right) \cdot \\
\cdot \sum\limits_{r=1}^{N} \mathop{T}\limits_{k=1}^{N}\left[\left(\mu_{A_{\mathrm{L}}^k}(\mathbf{v}) \overset{=}{*} 0\right)\cdot\left(\overline{z}_j^k \overset{\neq}{*} \overline{z}_j^r\right)+\left(\overline{z}_j^k \overset{=}{*} \overline{z}_j^r\right)\right] \\
-\sum\limits_{r=1}^{N} \overline{z}_j^r \cdot \mathop{T}\limits_{k=1}^{N}\left[\left(\mu_{A_{\mathrm{L}}^k}(\mathbf{v}) \overset{=}{*} 0\right)\cdot\left(\overline{z}_j^k \overset{\neq}{*} \overline{z}_j^r\right)+\left(\overline{z}_j^k \overset{=}{*} \overline{z}_j^r\right)\right] \cdot \\
\cdot\left(\begin{array}{l}
\sum\limits_{\substack{r=1 \\ r:\,\overline{z}_j^r=1}}^{N} \left.\dfrac{\partial \mathop{T}\limits_{k=1}^{N}\left[\left(\mu_{A_{\mathrm{L}}^k}(\mathbf{v}) \overset{=}{*} 0\right)\cdot\left(\overline{z}_j^k \overset{\neq}{*} \overline{z}_j^r\right)+\left(\overline{z}_j^k \overset{=}{*} \overline{z}_j^r\right)\right]}{\partial \mu_{A_{\mathrm{L}}^l}(\mathbf{v})}\right|_{l:\,\overline{z}_j^l=-1} \\
+\sum\limits_{\substack{r=1 \\ r:\,\overline{z}_j^r=-1}}^{N} \left.\dfrac{\partial \mathop{T}\limits_{k=1}^{N}\left[\left(\mu_{A_{\mathrm{L}}^k}(\mathbf{v}) \overset{=}{*} 0\right)\cdot\left(\overline{z}_j^k \overset{\neq}{*} \overline{z}_j^r\right)+\left(\overline{z}_j^k \overset{=}{*} \overline{z}_j^r\right)\right]}{\partial \mu_{A_{\mathrm{L}}^l}(\mathbf{v})}\right|_{l:\,\overline{z}_j^l=-1}
\end{array}\right)
\end{array}\right)}{\left[\sum\limits_{r=1}^{N} \mathop{T}\limits_{k=1}^{N}\left[\left(\mu_{A_{\mathrm{L}}^k}(\mathbf{v}) \overset{=}{*} 0\right)\cdot\left(\overline{z}_j^k \overset{\neq}{*} \overline{z}_j^r\right)+\left(\overline{z}_j^k \overset{=}{*} \overline{z}_j^r\right)\right]\right]^2}, \tag{8.151}
$$

$$\left.\frac{\partial \underline{\overline{z}}_j}{\partial \mu_{A_{\mathrm{L}}^l}(\mathbf{v})}\right|_{l:\,\overline{z}_j^l=-1}=\frac{\left(\begin{array}{c}\left(\displaystyle\sum_{\substack{r=1\\ r:\,\overline{z}_j^r=1}}^{N}\left.\frac{\partial \overset{N}{\underset{k=1}{T}}\left[\left(\mu_{A_{\mathrm{L}}^k}(\mathbf{v})\overset{=}{*}0\right)\cdot\left(\overline{z}_j^k\overset{\neq}{*}\overline{z}_j^r\right)+\left(\overline{z}_j^k\overset{=}{*}\overline{z}_j^r\right)\right]}{\partial \mu_{A_{\mathrm{L}}^l}(\mathbf{v})}\right|_{l:\,\overline{z}_j^l=-1}-0\right)\cdot\\ \cdot\displaystyle\sum_{r=1}^{N}\overset{N}{\underset{k=1}{T}}\left[\left(\mu_{A_{\mathrm{L}}^k}(\mathbf{v})\overset{=}{*}0\right)\cdot\left(\overline{z}_j^k\overset{\neq}{*}\overline{z}_j^r\right)+\left(\overline{z}_j^k\overset{=}{*}\overline{z}_j^r\right)\right]\\ -\displaystyle\sum_{r=1}^{N}\overline{z}_j^r\cdot\overset{N}{\underset{k=1}{T}}\left[\left(\mu_{A_{\mathrm{L}}^k}(\mathbf{v})\overset{=}{*}0\right)\cdot\left(\overline{z}_j^k\overset{\neq}{*}\overline{z}_j^r\right)+\left(\overline{z}_j^k\overset{=}{*}\overline{z}_j^r\right)\right]\cdot\\ \cdot\left(\displaystyle\sum_{\substack{r=1\\ r:\,\overline{z}_j^r=1}}^{N}\left.\frac{\partial \overset{N}{\underset{k=1}{T}}\left[\left(\mu_{A_{\mathrm{L}}^k}(\mathbf{v})\overset{=}{*}0\right)\cdot\left(\overline{z}_j^k\overset{\neq}{*}\overline{z}_j^r\right)+\left(\overline{z}_j^k\overset{=}{*}\overline{z}_j^r\right)\right]}{\partial \mu_{A_{\mathrm{L}}^l}(\mathbf{v})}\right|_{l:\,\overline{z}_j^l=-1}+0\right)\end{array}\right)}{\left[\displaystyle\sum_{r=1}^{N}\overset{N}{\underset{k=1}{T}}\left[\left(\mu_{A_{\mathrm{L}}^k}(\mathbf{v})\overset{=}{*}0\right)\cdot\left(\overline{z}_j^k\overset{\neq}{*}\overline{z}_j^r\right)+\left(\overline{z}_j^k\overset{=}{*}\overline{z}_j^r\right)\right]\right]^2},\tag{8.152}$$

$$\left.\frac{\partial \underline{\overline{z}}_j}{\partial \mu_{A_{\mathrm{L}}^l}(\mathbf{v})}\right|_{l:\,\overline{z}_j^l=-1}=\frac{\left(\begin{array}{c}\displaystyle\sum_{\substack{r=1\\ r:\,\overline{z}_j^r=1}}^{N}\left.\frac{\partial \overset{N}{\underset{k=1}{T}}\left[\left(\mu_{A_{\mathrm{L}}^k}(\mathbf{v})\overset{=}{*}0\right)\cdot\left(\overline{z}_j^k\overset{\neq}{*}\overline{z}_j^r\right)+\left(\overline{z}_j^k\overset{=}{*}\overline{z}_j^r\right)\right]}{\partial \mu_{A_{\mathrm{L}}^l}(\mathbf{v})}\right|_{l:\,\overline{z}_j^l=-1}\cdot\\ \cdot\displaystyle\sum_{r=1}^{N}\overset{N}{\underset{k=1}{T}}\left[\left(\mu_{A_{\mathrm{L}}^k}(\mathbf{v})\overset{=}{*}0\right)\cdot\left(\overline{z}_j^k\overset{\neq}{*}\overline{z}_j^r\right)+\left(\overline{z}_j^k\overset{=}{*}\overline{z}_j^r\right)\right]\\ -\displaystyle\sum_{r=1}^{N}\overline{z}_j^r\cdot\overset{N}{\underset{k=1}{T}}\left[\left(\mu_{A_{\mathrm{L}}^k}(\mathbf{v})\overset{=}{*}0\right)\cdot\left(\overline{z}_j^k\overset{\neq}{*}\overline{z}_j^r\right)+\left(\overline{z}_j^k\overset{=}{*}\overline{z}_j^r\right)\right]\cdot\\ \cdot\displaystyle\sum_{\substack{r=1\\ r:\,\overline{z}_j^r=1}}^{N}\left.\frac{\partial \overset{N}{\underset{k=1}{T}}\left[\left(\mu_{A_{\mathrm{L}}^k}(\mathbf{v})\overset{=}{*}0\right)\cdot\left(\overline{z}_j^k\overset{\neq}{*}\overline{z}_j^r\right)+\left(\overline{z}_j^k\overset{=}{*}\overline{z}_j^r\right)\right]}{\partial \mu_{A_{\mathrm{L}}^l}(\mathbf{v})}\right|_{l:\,\overline{z}_j^l=-1}\end{array}\right)}{\left[\displaystyle\sum_{r=1}^{N}\overset{N}{\underset{k=1}{T}}\left[\left(\mu_{A_{\mathrm{L}}^k}(\mathbf{v})\overset{=}{*}0\right)\cdot\left(\overline{z}_j^k\overset{\neq}{*}\overline{z}_j^r\right)+\left(\overline{z}_j^k\overset{=}{*}\overline{z}_j^r\right)\right]\right]^2},\tag{8.153}$$

$$\left.\frac{\partial \underline{\overline{z}}_j}{\partial \mu_{A_{\mathrm{L}}^l}(\mathbf{v})}\right|_{l:\,\overline{z}_j^l=-1}=\frac{\left(\begin{array}{c}\displaystyle\sum_{r=1}^{N}\left(1-\overline{z}_j^r\right)\cdot\overset{N}{\underset{k=1}{T}}\left[\left(\mu_{A_{\mathrm{L}}^k}(\mathbf{v})\overset{=}{*}0\right)\cdot\left(\overline{z}_j^k\overset{\neq}{*}\overline{z}_j^r\right)+\left(\overline{z}_j^k\overset{=}{*}\overline{z}_j^r\right)\right]\cdot\\ \cdot\displaystyle\sum_{\substack{r=1\\ r:\,\overline{z}_j^r=1}}^{N}\left.\frac{\partial \overset{N}{\underset{k=1}{T}}\left[\left(\mu_{A_{\mathrm{L}}^k}(\mathbf{v})\overset{=}{*}0\right)\cdot\left(\overline{z}_j^k\overset{\neq}{*}\overline{z}_j^r\right)+\left(\overline{z}_j^k\overset{=}{*}\overline{z}_j^r\right)\right]}{\partial \mu_{A_{\mathrm{L}}^l}(\mathbf{v})}\right|_{l:\,\overline{z}_j^l=-1}\end{array}\right)}{\left[\displaystyle\sum_{r=1}^{N}\overset{N}{\underset{k=1}{T}}\left[\left(\mu_{A_{\mathrm{L}}^k}(\mathbf{v})\overset{=}{*}0\right)\cdot\left(\overline{z}_j^k\overset{\neq}{*}\overline{z}_j^r\right)+\left(\overline{z}_j^k\overset{=}{*}\overline{z}_j^r\right)\right]\right]^2},\tag{8.154}$$

T–norm is nondecrasing, and operation $\left(\mu_{A_{\mathrm{L}}^k}(\mathbf{v})\overset{=}{*}0\right)$, which is a special case of negation, is nonincreasing, thus

$$\left.\frac{\partial \underline{\overline{z}}_j}{\partial \mu_{A_{\mathrm{L}}^l}(\mathbf{v})}\right|_{l:\,\overline{z}_j^l=-1}\leq 0,\tag{8.155}$$

In the second step for (8.148) and $l\colon \overline{z}_j^l = -1$:

$$\left.\frac{\partial \overline{\overline{z}}_j}{\partial \mu_{A_U^l}(\mathbf{v})}\right|_{l:\ \overline{z}_j^l=-1} = \left.\frac{\partial}{\partial \mu_{A_U^l}(\mathbf{v})} \frac{\sum\limits_{r=1}^{N} \overline{z}_j^r \cdot \mathop{T}\limits_{k=1}^{N} \left[\left(\mu_{A_U^k}(\mathbf{v}) \overset{=}{*} 0\right) \cdot \left(\overline{z}_j^k \overset{\neq}{*} \overline{z}_j^r\right) + \left(\overline{z}_j^k \overset{=}{*} \overline{z}_j^r\right)\right]}{\sum\limits_{r=1}^{N} \mathop{T}\limits_{k=1}^{N} \left[\left(\mu_{A_U^k}(\mathbf{v}) \overset{=}{*} 0\right) \cdot \left(\overline{z}_j^k \overset{\neq}{*} \overline{z}_j^r\right) + \left(\overline{z}_j^k \overset{=}{*} \overline{z}_j^r\right)\right]}\right|_{l:\ \overline{z}_j^l=-1}, \tag{8.156}$$

$$\left.\frac{\partial \overline{\overline{z}}_j}{\partial \mu_{A_U^l}(\mathbf{v})}\right|_{l:\ \overline{z}_j^l=-1} = \frac{\left(\begin{array}{c} \left.\frac{\partial \sum\limits_{r=1}^{N} \overline{z}_j^r \cdot \mathop{T}\limits_{k=1}^{N} \left[\left(\mu_{A_U^k}(\mathbf{v}) \overset{=}{*} 0\right) \cdot \left(\overline{z}_j^k \overset{\neq}{*} \overline{z}_j^r\right) + \left(\overline{z}_j^k \overset{=}{*} \overline{z}_j^r\right)\right]}{\partial \mu_{A_U^l}(\mathbf{v})}\right|_{l:\ \overline{z}_j^l=-1} \cdot \\ \cdot \sum\limits_{r=1}^{N} \mathop{T}\limits_{k=1}^{N} \left[\left(\mu_{A_U^k}(\mathbf{v}) \overset{=}{*} 0\right) \cdot \left(\overline{z}_j^k \overset{\neq}{*} \overline{z}_j^r\right) + \left(\overline{z}_j^k \overset{=}{*} \overline{z}_j^r\right)\right] \\ - \sum\limits_{r=1}^{N} \overline{z}_j^r \cdot \mathop{T}\limits_{k=1}^{N} \left[\left(\mu_{A_U^k}(\mathbf{v}) \overset{=}{*} 0\right) \cdot \left(\overline{z}_j^k \overset{\neq}{*} \overline{z}_j^r\right) + \left(\overline{z}_j^k \overset{=}{*} \overline{z}_j^r\right)\right] \\ \left.\frac{\partial \sum\limits_{r=1}^{N} \mathop{T}\limits_{k=1}^{N} \left[\left(\mu_{A_U^k}(\mathbf{v}) \overset{=}{*} 0\right) \cdot \left(\overline{z}_j^k \overset{\neq}{*} \overline{z}_j^r\right) + \left(\overline{z}_j^k \overset{=}{*} \overline{z}_j^r\right)\right]}{\partial \mu_{A_U^l}(\mathbf{v})}\right|_{l:\ \overline{z}_j^l=-1} \end{array}\right)}{\left[\sum\limits_{r=1}^{N} \mathop{T}\limits_{k=1}^{N} \left[\left(\mu_{A_U^k}(\mathbf{v}) \overset{=}{*} 0\right) \cdot \left(\overline{z}_j^k \overset{\neq}{*} \overline{z}_j^r\right) + \left(\overline{z}_j^k \overset{=}{*} \overline{z}_j^r\right)\right]\right]^2}, \tag{8.157}$$

$$\left.\frac{\partial \overline{\overline{z}}_j}{\partial \mu_{A_U^l}(\mathbf{v})}\right|_{l:\ \overline{z}_j^l=-1} = \frac{\left(\begin{array}{c} \left(\begin{array}{c} \sum\limits_{\substack{r=1\\ r:\ \overline{z}_j^r=1}}^{N} \left.\frac{\partial \mathop{T}\limits_{k=1}^{N} \left[\left(\mu_{A_U^k}(\mathbf{v}) \overset{=}{*} 0\right) \cdot \left(\overline{z}_j^k \overset{\neq}{*} \overline{z}_j^r\right) + \left(\overline{z}_j^k \overset{=}{*} \overline{z}_j^r\right)\right]}{\partial \mu_{A_U^l}(\mathbf{v})}\right|_{l:\ \overline{z}_j^l=-1} \\ - \sum\limits_{\substack{r=1\\ r:\ \overline{z}_j^r=-1}}^{N} \left.\frac{\partial \mathop{T}\limits_{k=1}^{N} \left[\left(\mu_{A_U^k}(\mathbf{v}) \overset{=}{*} 0\right) \cdot \left(\overline{z}_j^k \overset{\neq}{*} \overline{z}_j^r\right) + \left(\overline{z}_j^k \overset{=}{*} \overline{z}_j^r\right)\right]}{\partial \mu_{A_U^l}(\mathbf{v})}\right|_{l:\ \overline{z}_j^l=-1} \end{array}\right) \cdot \\ \cdot \sum\limits_{r=1}^{N} \mathop{T}\limits_{k=1}^{N} \left[\left(\mu_{A_U^k}(\mathbf{v}) \overset{=}{*} 0\right) \cdot \left(\overline{z}_j^k \overset{\neq}{*} \overline{z}_j^r\right) + \left(\overline{z}_j^k \overset{=}{*} \overline{z}_j^r\right)\right] \\ - \sum\limits_{r=1}^{N} \overline{z}_j^r \cdot \mathop{T}\limits_{k=1}^{N} \left[\left(\mu_{A_U^k}(\mathbf{v}) \overset{=}{*} 0\right) \cdot \left(\overline{z}_j^k \overset{\neq}{*} \overline{z}_j^r\right) + \left(\overline{z}_j^k \overset{=}{*} \overline{z}_j^r\right)\right] \cdot \\ \cdot \left(\begin{array}{c} \sum\limits_{\substack{r=1\\ r:\ \overline{z}_j^r=1}}^{N} \left.\frac{\partial \mathop{T}\limits_{k=1}^{N} \left[\left(\mu_{A_U^k}(\mathbf{v}) \overset{=}{*} 0\right) \cdot \left(\overline{z}_j^k \overset{\neq}{*} \overline{z}_j^r\right) + \left(\overline{z}_j^k \overset{=}{*} \overline{z}_j^r\right)\right]}{\partial \mu_{A_U^l}(\mathbf{v})}\right|_{l:\ \overline{z}_j^l=-1} \\ + \sum\limits_{\substack{r=1\\ r:\ \overline{z}_j^r=-1}}^{N} \left.\frac{\partial \mathop{T}\limits_{k=1}^{N} \left[\left(\mu_{A_U^k}(\mathbf{v}) \overset{=}{*} 0\right) \cdot \left(\overline{z}_j^k \overset{\neq}{*} \overline{z}_j^r\right) + \left(\overline{z}_j^k \overset{=}{*} \overline{z}_j^r\right)\right]}{\partial \mu_{A_U^l}(\mathbf{v})}\right|_{l:\ \overline{z}_j^l=-1} \end{array}\right) \end{array}\right)}{\left[\sum\limits_{r=1}^{N} \mathop{T}\limits_{k=1}^{N} \left[\left(\mu_{A_U^k}(\mathbf{v}) \overset{=}{*} 0\right) \cdot \left(\overline{z}_j^k \overset{\neq}{*} \overline{z}_j^r\right) + \left(\overline{z}_j^k \overset{=}{*} \overline{z}_j^r\right)\right]\right]^2}, \tag{8.158}$$

$$\left.\frac{\partial \overline{\overline{z}}_j}{\partial \mu_{A_{\mathrm{U}}^{l}}(\mathbf{v})}\right|_{l:\,\overline{z}_j^l=-1} = \frac{\begin{pmatrix} \left(\sum\limits_{\substack{r=1\\ r:\,\overline{z}_j^r=1}}^{N} \left.\frac{\partial \mathop{T}\limits_{k=1}^{N}\left[\left(\mu_{A_{\mathrm{U}}^{k}}(\mathbf{v}) \stackrel{=}{*} 0\right)\cdot\left(\overline{z}_j^k \stackrel{\neq}{*} \overline{z}_j^r\right)+\left(\overline{z}_j^k \stackrel{=}{*} \overline{z}_j^r\right)\right]}{\partial \mu_{A_{\mathrm{U}}^{l}}(\mathbf{v})}\right|_{l:\,\overline{z}_j^l=-1} - 0 \right) \cdot \\ \cdot \sum\limits_{r=1}^{N} \mathop{T}\limits_{k=1}^{N}\left[\left(\mu_{A_{\mathrm{U}}^{k}}(\mathbf{v}) \stackrel{=}{*} 0\right)\cdot\left(\overline{z}_j^k \stackrel{\neq}{*} \overline{z}_j^r\right)+\left(\overline{z}_j^k \stackrel{=}{*} \overline{z}_j^r\right)\right] \\ - \sum\limits_{r=1}^{N} \overline{z}_j^r \cdot \mathop{T}\limits_{k=1}^{N}\left[\left(\mu_{A_{\mathrm{U}}^{k}}(\mathbf{v}) \stackrel{=}{*} 0\right)\cdot\left(\overline{z}_j^k \stackrel{\neq}{*} \overline{z}_j^r\right)+\left(\overline{z}_j^k \stackrel{=}{*} \overline{z}_j^r\right)\right] \cdot \\ \cdot \left(\sum\limits_{\substack{r=1\\ r:\,\overline{z}_j^r=1}}^{N} \left.\frac{\partial \mathop{T}\limits_{k=1}^{N}\left[\left(\mu_{A_{\mathrm{U}}^{k}}(\mathbf{v}) \stackrel{=}{*} 0\right)\cdot\left(\overline{z}_j^k \stackrel{\neq}{*} \overline{z}_j^r\right)+\left(\overline{z}_j^k \stackrel{=}{*} \overline{z}_j^r\right)\right]}{\partial \mu_{A_{\mathrm{U}}^{l}}(\mathbf{v})}\right|_{l:\,\overline{z}_j^l=-1} + 0 \right) \end{pmatrix}}{\left[\sum\limits_{r=1}^{N} \mathop{T}\limits_{k=1}^{N}\left[\left(\mu_{A_{\mathrm{U}}^{k}}(\mathbf{v}) \stackrel{=}{*} 0\right)\cdot\left(\overline{z}_j^k \stackrel{\neq}{*} \overline{z}_j^r\right)+\left(\overline{z}_j^k \stackrel{=}{*} \overline{z}_j^r\right)\right]\right]^2}, \tag{8.159}$$

$$\left.\frac{\partial \overline{\overline{z}}_j}{\partial \mu_{A_{\mathrm{U}}^{l}}(\mathbf{v})}\right|_{l:\,\overline{z}_j^l=-1} = \frac{\begin{pmatrix} \sum\limits_{\substack{r=1\\ r:\,\overline{z}_j^r=1}}^{N} \left.\frac{\partial \mathop{T}\limits_{k=1}^{N}\left[\left(\mu_{A_{\mathrm{U}}^{k}}(\mathbf{v}) \stackrel{=}{*} 0\right)\cdot\left(\overline{z}_j^k \stackrel{\neq}{*} \overline{z}_j^r\right)+\left(\overline{z}_j^k \stackrel{=}{*} \overline{z}_j^r\right)\right]}{\partial \mu_{A_{\mathrm{U}}^{l}}(\mathbf{v})}\right|_{l:\,\overline{z}_j^l=-1} \cdot \\ \cdot \sum\limits_{r=1}^{N} \mathop{T}\limits_{k=1}^{N}\left[\left(\mu_{A_{\mathrm{U}}^{k}}(\mathbf{v}) \stackrel{=}{*} 0\right)\cdot\left(\overline{z}_j^k \stackrel{\neq}{*} \overline{z}_j^r\right)+\left(\overline{z}_j^k \stackrel{=}{*} \overline{z}_j^r\right)\right] \\ - \sum\limits_{r=1}^{N} \overline{z}_j^r \cdot \mathop{T}\limits_{k=1}^{N}\left[\left(\mu_{A_{\mathrm{U}}^{k}}(\mathbf{v}) \stackrel{=}{*} 0\right)\cdot\left(\overline{z}_j^k \stackrel{\neq}{*} \overline{z}_j^r\right)+\left(\overline{z}_j^k \stackrel{=}{*} \overline{z}_j^r\right)\right] \cdot \\ \cdot \sum\limits_{\substack{r=1\\ r:\,\overline{z}_j^r=1}}^{N} \left.\frac{\partial \mathop{T}\limits_{k=1}^{N}\left[\left(\mu_{A_{\mathrm{U}}^{k}}(\mathbf{v}) \stackrel{=}{*} 0\right)\cdot\left(\overline{z}_j^k \stackrel{\neq}{*} \overline{z}_j^r\right)+\left(\overline{z}_j^k \stackrel{=}{*} \overline{z}_j^r\right)\right]}{\partial \mu_{A_{\mathrm{U}}^{l}}(\mathbf{v})}\right|_{l:\,\overline{z}_j^l=-1} \end{pmatrix}}{\left[\sum\limits_{r=1}^{N} \mathop{T}\limits_{k=1}^{N}\left[\left(\mu_{A_{\mathrm{U}}^{k}}(\mathbf{v}) \stackrel{=}{*} 0\right)\cdot\left(\overline{z}_j^k \stackrel{\neq}{*} \overline{z}_j^r\right)+\left(\overline{z}_j^k \stackrel{=}{*} \overline{z}_j^r\right)\right]\right]^2}, \tag{8.160}$$

$$\left.\frac{\partial \overline{\overline{z}}_j}{\partial \mu_{A_{\mathrm{U}}^{l}}(\mathbf{v})}\right|_{l:\,\overline{z}_j^l=-1} = \frac{\begin{pmatrix} \sum\limits_{r=1}^{N} \left(1-\overline{z}_j^r\right) \cdot \mathop{T}\limits_{k=1}^{N}\left[\left(\mu_{A_{\mathrm{U}}^{k}}(\mathbf{v}) \stackrel{=}{*} 0\right)\cdot\left(\overline{z}_j^k \stackrel{\neq}{*} \overline{z}_j^r\right)+\left(\overline{z}_j^k \stackrel{=}{*} \overline{z}_j^r\right)\right] \cdot \\ \cdot \sum\limits_{\substack{r=1\\ r:\,\overline{z}_j^r=1}}^{N} \left.\frac{\partial \mathop{T}\limits_{k=1}^{N}\left[\left(\mu_{A_{\mathrm{U}}^{k}}(\mathbf{v}) \stackrel{=}{*} 0\right)\cdot\left(\overline{z}_j^k \stackrel{\neq}{*} \overline{z}_j^r\right)+\left(\overline{z}_j^k \stackrel{=}{*} \overline{z}_j^r\right)\right]}{\partial \mu_{A_{\mathrm{U}}^{l}}(\mathbf{v})}\right|_{l:\,\overline{z}_j^l=-1} \end{pmatrix}}{\left[\sum\limits_{r=1}^{N} \mathop{T}\limits_{k=1}^{N}\left[\left(\mu_{A_{\mathrm{U}}^{k}}(\mathbf{v}) \stackrel{=}{*} 0\right)\cdot\left(\overline{z}_j^k \stackrel{\neq}{*} \overline{z}_j^r\right)+\left(\overline{z}_j^k \stackrel{=}{*} \overline{z}_j^r\right)\right]\right]^2}, \tag{8.161}$$

T–norm is nondecrasing, and operation $\left(\mu_{A_{\mathrm{U}}^{k}}(\mathbf{v}) \stackrel{=}{*} 0\right)$, which is a special case of negation, is nonincreasing, thus

$$\left.\frac{\partial \overline{\overline{z}}_j}{\partial \mu_{A_{\mathrm{U}}^{l}}(\mathbf{v})}\right|_{l:\,\overline{z}_j^l=-1} \leq 0, \tag{8.162}$$

In the third step for (8.147) and $l\colon \overline{\overline{z}}_j^l = 1$:

$$\left.\frac{\partial \underline{\overline{z}}_j}{\partial \mu_{A_{\mathrm{L}}^l}(\mathbf{v})}\right|_{l:\, \overline{\overline{z}}_j^l=-1} = \frac{\partial}{\partial \mu_{A_{\mathrm{L}}^l}(\mathbf{v})} \left.\frac{\sum\limits_{r=1}^{N} \overline{\overline{z}}_j^r \cdot \mathop{T}\limits_{k=1}^{N} \left[\left(\mu_{A_{\mathrm{L}}^k}(\mathbf{v}) \overset{=}{*} 0\right) \cdot \left(\overline{\overline{z}}_j^k \overset{\neq}{*} \overline{\overline{z}}_j^r\right) + \left(\overline{\overline{z}}_j^k \overset{=}{*} \overline{\overline{z}}_j^r\right)\right]}{\sum\limits_{r=1}^{N} \mathop{T}\limits_{k=1}^{N} \left[\left(\mu_{A_{\mathrm{L}}^k}(\mathbf{v}) \overset{=}{*} 0\right) \cdot \left(\overline{\overline{z}}_j^k \overset{\neq}{*} \overline{\overline{z}}_j^r\right) + \left(\overline{\overline{z}}_j^k \overset{=}{*} \overline{\overline{z}}_j^r\right)\right]}\right|_{l:\, \overline{\overline{z}}_j^l=1}, \tag{8.163}$$

$$\left.\frac{\partial \underline{\overline{z}}_j}{\partial \mu_{A_{\mathrm{L}}^l}(\mathbf{v})}\right|_{l:\, \overline{\overline{z}}_j^l=-1} = \frac{\left(\begin{array}{c} \left.\frac{\partial \sum\limits_{r=1}^{N} \overline{\overline{z}}_j^r \cdot \mathop{T}\limits_{k=1}^{N} \left[\left(\mu_{A_{\mathrm{L}}^k}(\mathbf{v}) \overset{=}{*} 0\right) \cdot \left(\overline{\overline{z}}_j^k \overset{\neq}{*} \overline{\overline{z}}_j^r\right) + \left(\overline{\overline{z}}_j^k \overset{=}{*} \overline{\overline{z}}_j^r\right)\right]}{\partial \mu_{A_{\mathrm{L}}^l}(\mathbf{v})}\right|_{l:\, \overline{\overline{z}}_j^l=1} \cdot \\ \cdot \sum\limits_{r=1}^{N} \mathop{T}\limits_{k=1}^{N} \left[\left(\mu_{A_{\mathrm{L}}^k}(\mathbf{v}) \overset{=}{*} 0\right) \cdot \left(\overline{\overline{z}}_j^k \overset{\neq}{*} \overline{\overline{z}}_j^r\right) + \left(\overline{\overline{z}}_j^k \overset{=}{*} \overline{\overline{z}}_j^r\right)\right] \\ - \sum\limits_{r=1}^{N} \overline{\overline{z}}_j^r \cdot \mathop{T}\limits_{k=1}^{N} \left[\left(\mu_{A_{\mathrm{L}}^k}(\mathbf{v}) \overset{=}{*} 0\right) \cdot \left(\overline{\overline{z}}_j^k \overset{\neq}{*} \overline{\overline{z}}_j^r\right) + \left(\overline{\overline{z}}_j^k \overset{=}{*} \overline{\overline{z}}_j^r\right)\right] \\ \left.\frac{\partial \sum\limits_{r=1}^{N} \mathop{T}\limits_{k=1}^{N} \left[\left(\mu_{A_{\mathrm{L}}^k}(\mathbf{v}) \overset{=}{*} 0\right) \cdot \left(\overline{\overline{z}}_j^k \overset{\neq}{*} \overline{\overline{z}}_j^r\right) + \left(\overline{\overline{z}}_j^k \overset{=}{*} \overline{\overline{z}}_j^r\right)\right]}{\partial \mu_{A_{\mathrm{L}}^l}(\mathbf{v})}\right|_{l:\, \overline{\overline{z}}_j^l=1} \end{array}\right)}{\left[\sum\limits_{r=1}^{N} \mathop{T}\limits_{k=1}^{N} \left[\left(\mu_{A_{\mathrm{L}}^k}(\mathbf{v}) \overset{=}{*} 0\right) \cdot \left(\overline{\overline{z}}_j^k \overset{\neq}{*} \overline{\overline{z}}_j^r\right) + \left(\overline{\overline{z}}_j^k \overset{=}{*} \overline{\overline{z}}_j^r\right)\right]\right]^2}, \tag{8.164}$$

$$\left.\frac{\partial \underline{\overline{z}}_j}{\partial \mu_{A_{\mathrm{L}}^l}(\mathbf{v})}\right|_{l:\, \overline{\overline{z}}_j^l=-1} = \frac{\left(\begin{array}{c} \left(\begin{array}{c} \sum\limits_{\substack{r=1\\ r:\, \overline{\overline{z}}_j^r=1}}^{N} \left.\frac{\partial \mathop{T}\limits_{k=1}^{N} \left[\left(\mu_{A_{\mathrm{L}}^k}(\mathbf{v}) \overset{=}{*} 0\right) \cdot \left(\overline{\overline{z}}_j^k \overset{\neq}{*} \overline{\overline{z}}_j^r\right) + \left(\overline{\overline{z}}_j^k \overset{=}{*} \overline{\overline{z}}_j^r\right)\right]}{\partial \mu_{A_{\mathrm{L}}^l}(\mathbf{v})}\right|_{l:\, \overline{\overline{z}}_j^l=1} \\ - \sum\limits_{\substack{r=1\\ r:\, \overline{\overline{z}}_j^r=-1}}^{N} \left.\frac{\partial \mathop{T}\limits_{k=1}^{N} \left[\left(\mu_{A_{\mathrm{L}}^k}(\mathbf{v}) \overset{=}{*} 0\right) \cdot \left(\overline{\overline{z}}_j^k \overset{\neq}{*} \overline{\overline{z}}_j^r\right) + \left(\overline{\overline{z}}_j^k \overset{=}{*} \overline{\overline{z}}_j^r\right)\right]}{\partial \mu_{A_{\mathrm{L}}^l}(\mathbf{v})}\right|_{l:\, \overline{\overline{z}}_j^l=1} \end{array}\right) \cdot \\ \cdot \sum\limits_{r=1}^{N} \mathop{T}\limits_{k=1}^{N} \left[\left(\mu_{A_{\mathrm{L}}^k}(\mathbf{v}) \overset{=}{*} 0\right) \cdot \left(\overline{\overline{z}}_j^k \overset{\neq}{*} \overline{\overline{z}}_j^r\right) + \left(\overline{\overline{z}}_j^k \overset{=}{*} \overline{\overline{z}}_j^r\right)\right] \\ - \sum\limits_{r=1}^{N} \overline{\overline{z}}_j^r \cdot \mathop{T}\limits_{k=1}^{N} \left[\left(\mu_{A_{\mathrm{L}}^k}(\mathbf{v}) \overset{=}{*} 0\right) \cdot \left(\overline{\overline{z}}_j^k \overset{\neq}{*} \overline{\overline{z}}_j^r\right) + \left(\overline{\overline{z}}_j^k \overset{=}{*} \overline{\overline{z}}_j^r\right)\right] \cdot \\ \cdot \left(\begin{array}{c} \sum\limits_{\substack{r=1\\ r:\, \overline{\overline{z}}_j^r=1}}^{N} \left.\frac{\partial \mathop{T}\limits_{k=1}^{N} \left[\left(\mu_{A_{\mathrm{L}}^k}(\mathbf{v}) \overset{=}{*} 0\right) \cdot \left(\overline{\overline{z}}_j^k \overset{\neq}{*} \overline{\overline{z}}_j^r\right) + \left(\overline{\overline{z}}_j^k \overset{=}{*} \overline{\overline{z}}_j^r\right)\right]}{\partial \mu_{A_{\mathrm{L}}^l}(\mathbf{v})}\right|_{l:\, \overline{\overline{z}}_j^l=1} \\ + \sum\limits_{\substack{r=1\\ r:\, \overline{\overline{z}}_j^r=-1}}^{N} \left.\frac{\partial \mathop{T}\limits_{k=1}^{N} \left[\left(\mu_{A_{\mathrm{L}}^k}(\mathbf{v}) \overset{=}{*} 0\right) \cdot \left(\overline{\overline{z}}_j^k \overset{\neq}{*} \overline{\overline{z}}_j^r\right) + \left(\overline{\overline{z}}_j^k \overset{=}{*} \overline{\overline{z}}_j^r\right)\right]}{\partial \mu_{A_{\mathrm{L}}^l}(\mathbf{v})}\right|_{l:\, \overline{\overline{z}}_j^l=1} \end{array}\right) \end{array}\right)}{\left[\sum\limits_{r=1}^{N} \mathop{T}\limits_{k=1}^{N} \left[\left(\mu_{A_{\mathrm{L}}^k}(\mathbf{v}) \overset{=}{*} 0\right) \cdot \left(\overline{\overline{z}}_j^k \overset{\neq}{*} \overline{\overline{z}}_j^r\right) + \left(\overline{\overline{z}}_j^k \overset{=}{*} \overline{\overline{z}}_j^r\right)\right]\right]^2}, \tag{8.165}$$

$$\left.\frac{\partial \underline{\overline{z}}_j}{\partial \mu_{A^l_{\mathrm{L}}}(\mathbf{v})}\right|_{l:\,\overline{z}^l_j=-1} = \frac{\left(\begin{array}{l}\left(0-\sum\limits_{\substack{r=1\\ r:\,\overline{z}^r_j=-1}}^{N}\left.\frac{\partial \mathop{T}\limits_{k=1}^{N}\left[\left(\mu_{A^k_{\mathrm{L}}}(\mathbf{v})\overset{=}{*}0\right)\cdot\left(\overline{z}^k_j\overset{\neq}{*}\overline{z}^r_j\right)+\left(\overline{z}^k_j\overset{=}{*}\overline{z}^r_j\right)\right]}{\partial \mu_{A^l_{\mathrm{L}}}(\mathbf{v})}\right|_{l:\,\overline{z}^l_j=1}\right)\cdot\\ \cdot\sum\limits_{r=1}^{N}\mathop{T}\limits_{k=1}^{N}\left[\left(\mu_{A^k_{\mathrm{L}}}(\mathbf{v})\overset{=}{*}0\right)\cdot\left(\overline{z}^k_j\overset{\neq}{*}\overline{z}^r_j\right)+\left(\overline{z}^k_j\overset{=}{*}\overline{z}^r_j\right)\right]\\ -\sum\limits_{r=1}^{N}\overline{z}^r_j\cdot\mathop{T}\limits_{k=1}^{N}\left[\left(\mu_{A^k_{\mathrm{L}}}(\mathbf{v})\overset{=}{*}0\right)\cdot\left(\overline{z}^k_j\overset{\neq}{*}\overline{z}^r_j\right)+\left(\overline{z}^k_j\overset{=}{*}\overline{z}^r_j\right)\right]\cdot\\ \cdot\left(0+\sum\limits_{\substack{r=1\\ r:\,\overline{z}^r_j=-1}}^{N}\left.\frac{\partial \mathop{T}\limits_{k=1}^{N}\left[\left(\mu_{A^k_{\mathrm{L}}}(\mathbf{v})\overset{=}{*}0\right)\cdot\left(\overline{z}^k_j\overset{\neq}{*}\overline{z}^r_j\right)+\left(\overline{z}^k_j\overset{=}{*}\overline{z}^r_j\right)\right]}{\partial \mu_{A^l_{\mathrm{L}}}(\mathbf{v})}\right|_{l:\,\overline{z}^l_j=1}\right)\end{array}\right)}{\left[\sum\limits_{r=1}^{N}\mathop{T}\limits_{k=1}^{N}\left[\left(\mu_{A^k_{\mathrm{L}}}(\mathbf{v})\overset{=}{*}0\right)\cdot\left(\overline{z}^k_j\overset{\neq}{*}\overline{z}^r_j\right)+\left(\overline{z}^k_j\overset{=}{*}\overline{z}^r_j\right)\right]\right]^2}, \tag{8.166}$$

$$\left.\frac{\partial \underline{\overline{z}}_j}{\partial \mu_{A^l_{\mathrm{L}}}(\mathbf{v})}\right|_{l:\,\overline{z}^l_j=-1} = \frac{\left(\begin{array}{l}-\sum\limits_{\substack{r=1\\ r:\,\overline{z}^r_j=-1}}^{N}\left.\frac{\partial \mathop{T}\limits_{k=1}^{N}\left[\left(\mu_{A^k_{\mathrm{L}}}(\mathbf{v})\overset{=}{*}0\right)\cdot\left(\overline{z}^k_j\overset{\neq}{*}\overline{z}^r_j\right)+\left(\overline{z}^k_j\overset{=}{*}\overline{z}^r_j\right)\right]}{\partial \mu_{A^l_{\mathrm{L}}}(\mathbf{v})}\right|_{l:\,\overline{z}^l_j=1}\cdot\\ \cdot\sum\limits_{r=1}^{N}\mathop{T}\limits_{k=1}^{N}\left[\left(\mu_{A^k_{\mathrm{L}}}(\mathbf{v})\overset{=}{*}0\right)\cdot\left(\overline{z}^k_j\overset{\neq}{*}\overline{z}^r_j\right)+\left(\overline{z}^k_j\overset{=}{*}\overline{z}^r_j\right)\right]\\ -\sum\limits_{r=1}^{N}\overline{z}^r_j\cdot\mathop{T}\limits_{k=1}^{N}\left[\left(\mu_{A^k_{\mathrm{L}}}(\mathbf{v})\overset{=}{*}0\right)\cdot\left(\overline{z}^k_j\overset{\neq}{*}\overline{z}^r_j\right)+\left(\overline{z}^k_j\overset{=}{*}\overline{z}^r_j\right)\right]\cdot\\ \cdot\sum\limits_{\substack{r=1\\ r:\,\overline{z}^r_j=-1}}^{N}\left.\frac{\partial \mathop{T}\limits_{k=1}^{N}\left[\left(\mu_{A^k_{\mathrm{L}}}(\mathbf{v})\overset{=}{*}0\right)\cdot\left(\overline{z}^k_j\overset{\neq}{*}\overline{z}^r_j\right)+\left(\overline{z}^k_j\overset{=}{*}\overline{z}^r_j\right)\right]}{\partial \mu_{A^l_{\mathrm{L}}}(\mathbf{v})}\right|_{l:\,\overline{z}^l_j=1}\end{array}\right)}{\left[\sum\limits_{r=1}^{N}\mathop{T}\limits_{k=1}^{N}\left[\left(\mu_{A^k_{\mathrm{L}}}(\mathbf{v})\overset{=}{*}0\right)\cdot\left(\overline{z}^k_j\overset{\neq}{*}\overline{z}^r_j\right)+\left(\overline{z}^k_j\overset{=}{*}\overline{z}^r_j\right)\right]\right]^2}, \tag{8.167}$$

$$\left.\frac{\partial \underline{\overline{z}}_j}{\partial \mu_{A^l_{\mathrm{L}}}(\mathbf{v})}\right|_{l:\,\overline{z}^l_j=-1} = -\frac{\left(\begin{array}{l}\sum\limits_{r=1}^{N}\left(1+\overline{z}^r_j\right)\cdot\mathop{T}\limits_{k=1}^{N}\left[\left(\mu_{A^k_{\mathrm{L}}}(\mathbf{v})\overset{=}{*}0\right)\cdot\left(\overline{z}^k_j\overset{\neq}{*}\overline{z}^r_j\right)+\left(\overline{z}^k_j\overset{=}{*}\overline{z}^r_j\right)\right]\cdot\\ \cdot\sum\limits_{\substack{r=1\\ r:\,\overline{z}^r_j=-1}}^{N}\left.\frac{\partial \mathop{T}\limits_{k=1}^{N}\left[\left(\mu_{A^k_{\mathrm{L}}}(\mathbf{v})\overset{=}{*}0\right)\cdot\left(\overline{z}^k_j\overset{\neq}{*}\overline{z}^r_j\right)+\left(\overline{z}^k_j\overset{=}{*}\overline{z}^r_j\right)\right]}{\partial \mu_{A^l_{\mathrm{L}}}(\mathbf{v})}\right|_{l:\,\overline{z}^l_j=1}\end{array}\right)}{\left[\sum\limits_{r=1}^{N}\mathop{T}\limits_{k=1}^{N}\left[\left(\mu_{A^k_{\mathrm{L}}}(\mathbf{v})\overset{=}{*}0\right)\cdot\left(\overline{z}^k_j\overset{\neq}{*}\overline{z}^r_j\right)+\left(\overline{z}^k_j\overset{=}{*}\overline{z}^r_j\right)\right]\right]^2}, \tag{8.168}$$

T–norm is nondecrasing, and operation $\left(\mu_{A^k_{\mathrm{U}}}(\mathbf{v})\overset{=}{*}0\right)$, which is a special case of negation, is nonincreasing, thus

$$\left.\frac{\partial \underline{\overline{z}}_j}{\partial \mu_{A^l_{\mathrm{L}}}(\mathbf{v})}\right|_{l:\,\overline{z}^l_j=1} \geq 0, \tag{8.169}$$

In the fourth step, for (8.148) and $l\colon \overline{z}_j^l = 1$:

$$\left.\frac{\partial \overline{\overline{z}}_j}{\partial \mu_{A_{\mathrm{U}}^l}(\mathbf{v})}\right|_{l\colon \overline{z}_j^l=-1} = \left.\frac{\partial}{\partial \mu_{A_{\mathrm{U}}^l}(\mathbf{v})} \frac{\sum\limits_{r=1}^{N} \overline{z}_j^r \cdot \mathop{T}\limits_{k=1}^{N}\left[\left(\mu_{A_{\mathrm{U}}^k}(\mathbf{v}) \overset{=}{*} 0\right)\cdot\left(\overline{z}_j^k \overset{\neq}{*} \overline{z}_j^r\right)+\left(\overline{z}_j^k \overset{=}{*} \overline{z}_j^r\right)\right]}{\sum\limits_{r=1}^{N} \mathop{T}\limits_{k=1}^{N}\left[\left(\mu_{A_{\mathrm{U}}^k}(\mathbf{v}) \overset{=}{*} 0\right)\cdot\left(\overline{z}_j^k \overset{\neq}{*} \overline{z}_j^r\right)+\left(\overline{z}_j^k \overset{=}{*} \overline{z}_j^r\right)\right]}\right|_{l\colon \overline{z}_j^l=1}, \tag{8.170}$$

$$\left.\frac{\partial \overline{\overline{z}}_j}{\partial \mu_{A_{\mathrm{U}}^l}(\mathbf{v})}\right|_{l\colon \overline{z}_j^l=-1} = \frac{\left(\begin{array}{c} \left.\frac{\partial \sum\limits_{r=1}^{N} \overline{z}_j^r \cdot \mathop{T}\limits_{k=1}^{N}\left[\left(\mu_{A_{\mathrm{U}}^k}(\mathbf{v}) \overset{=}{*} 0\right)\cdot\left(\overline{z}_j^k \overset{\neq}{*} \overline{z}_j^r\right)+\left(\overline{z}_j^k \overset{=}{*} \overline{z}_j^r\right)\right]}{\partial \mu_{A_{\mathrm{U}}^l}(\mathbf{v})}\right|_{l\colon \overline{z}_j^l=1} \cdot \\ \cdot \sum\limits_{r=1}^{N} \mathop{T}\limits_{k=1}^{N}\left[\left(\mu_{A_{\mathrm{U}}^k}(\mathbf{v}) \overset{=}{*} 0\right)\cdot\left(\overline{z}_j^k \overset{\neq}{*} \overline{z}_j^r\right)+\left(\overline{z}_j^k \overset{=}{*} \overline{z}_j^r\right)\right] \\ -\sum\limits_{r=1}^{N} \overline{z}_j^r \cdot \mathop{T}\limits_{k=1}^{N}\left[\left(\mu_{A_{\mathrm{U}}^k}(\mathbf{v}) \overset{=}{*} 0\right)\cdot\left(\overline{z}_j^k \overset{\neq}{*} \overline{z}_j^r\right)+\left(\overline{z}_j^k \overset{=}{*} \overline{z}_j^r\right)\right] \\ \left.\frac{\partial \sum\limits_{r=1}^{N} \mathop{T}\limits_{k=1}^{N}\left[\left(\mu_{A_{\mathrm{U}}^k}(\mathbf{v}) \overset{=}{*} 0\right)\cdot\left(\overline{z}_j^k \overset{\neq}{*} \overline{z}_j^r\right)+\left(\overline{z}_j^k \overset{=}{*} \overline{z}_j^r\right)\right]}{\partial \mu_{A_{\mathrm{U}}^l}(\mathbf{v})}\right|_{l\colon \overline{z}_j^l=1} \end{array}\right)}{\left[\sum\limits_{r=1}^{N} \mathop{T}\limits_{k=1}^{N}\left[\left(\mu_{A_{\mathrm{U}}^k}(\mathbf{v}) \overset{=}{*} 0\right)\cdot\left(\overline{z}_j^k \overset{\neq}{*} \overline{z}_j^r\right)+\left(\overline{z}_j^k \overset{=}{*} \overline{z}_j^r\right)\right]\right]^2}, \tag{8.171}$$

$$\left.\frac{\partial \overline{\overline{z}}_j}{\partial \mu_{A_{\mathrm{U}}^l}(\mathbf{v})}\right|_{l\colon \overline{z}_j^l=-1} = \frac{\left(\begin{array}{c} \left(\begin{array}{c} \sum\limits_{\substack{r=1 \\ r\colon \overline{z}_j^r=1}}^{N} \left.\frac{\partial \mathop{T}\limits_{k=1}^{N}\left[\left(\mu_{A_{\mathrm{U}}^k}(\mathbf{v}) \overset{=}{*} 0\right)\cdot\left(\overline{z}_j^k \overset{\neq}{*} \overline{z}_j^r\right)+\left(\overline{z}_j^k \overset{=}{*} \overline{z}_j^r\right)\right]}{\partial \mu_{A_{\mathrm{U}}^l}(\mathbf{v})}\right|_{l\colon \overline{z}_j^l=1} \\ -\sum\limits_{\substack{r=1 \\ r\colon \overline{z}_j^r=-1}}^{N} \left.\frac{\partial \mathop{T}\limits_{k=1}^{N}\left[\left(\mu_{A_{\mathrm{U}}^k}(\mathbf{v}) \overset{=}{*} 0\right)\cdot\left(\overline{z}_j^k \overset{\neq}{*} \overline{z}_j^r\right)+\left(\overline{z}_j^k \overset{=}{*} \overline{z}_j^r\right)\right]}{\partial \mu_{A_{\mathrm{U}}^l}(\mathbf{v})}\right|_{l\colon \overline{z}_j^l=1} \end{array}\right) \cdot \\ \cdot \sum\limits_{r=1}^{N} \mathop{T}\limits_{k=1}^{N}\left[\left(\mu_{A_{\mathrm{U}}^k}(\mathbf{v}) \overset{=}{*} 0\right)\cdot\left(\overline{z}_j^k \overset{\neq}{*} \overline{z}_j^r\right)+\left(\overline{z}_j^k \overset{=}{*} \overline{z}_j^r\right)\right] \\ -\sum\limits_{r=1}^{N} \overline{z}_j^r \cdot \mathop{T}\limits_{k=1}^{N}\left[\left(\mu_{A_{\mathrm{U}}^k}(\mathbf{v}) \overset{=}{*} 0\right)\cdot\left(\overline{z}_j^k \overset{\neq}{*} \overline{z}_j^r\right)+\left(\overline{z}_j^k \overset{=}{*} \overline{z}_j^r\right)\right] \cdot \\ \cdot \left(\begin{array}{c} \sum\limits_{\substack{r=1 \\ r\colon \overline{z}_j^r=1}}^{N} \left.\frac{\partial \mathop{T}\limits_{k=1}^{N}\left[\left(\mu_{A_{\mathrm{U}}^k}(\mathbf{v}) \overset{=}{*} 0\right)\cdot\left(\overline{z}_j^k \overset{\neq}{*} \overline{z}_j^r\right)+\left(\overline{z}_j^k \overset{=}{*} \overline{z}_j^r\right)\right]}{\partial \mu_{A_{\mathrm{U}}^l}(\mathbf{v})}\right|_{l\colon \overline{z}_j^l=1} \\ +\sum\limits_{\substack{r=1 \\ r\colon \overline{z}_j^r=-1}}^{N} \left.\frac{\partial \mathop{T}\limits_{k=1}^{N}\left[\left(\mu_{A_{\mathrm{U}}^k}(\mathbf{v}) \overset{=}{*} 0\right)\cdot\left(\overline{z}_j^k \overset{\neq}{*} \overline{z}_j^r\right)+\left(\overline{z}_j^k \overset{=}{*} \overline{z}_j^r\right)\right]}{\partial \mu_{A_{\mathrm{U}}^l}(\mathbf{v})}\right|_{l\colon \overline{z}_j^l=1} \end{array}\right) \end{array}\right)}{\left[\sum\limits_{r=1}^{N} \mathop{T}\limits_{k=1}^{N}\left[\left(\mu_{A_{\mathrm{U}}^k}(\mathbf{v}) \overset{=}{*} 0\right)\cdot\left(\overline{z}_j^k \overset{\neq}{*} \overline{z}_j^r\right)+\left(\overline{z}_j^k \overset{=}{*} \overline{z}_j^r\right)\right]\right]^2}, \tag{8.172}$$

$$\left.\frac{\partial \overline{z}_j}{\partial \mu_{A_{\mathrm{U}}^l}(\mathbf{v})}\right|_{l:\,\overline{z}_j^l=-1} = \frac{\begin{pmatrix} \left(0-\sum\limits_{\substack{r=1\\ r:\,\overline{z}_j^r=-1}}^{N} \left.\frac{\partial \mathop{T}\limits_{k=1}^{N}\left[\left(\mu_{A_{\mathrm{U}}^k}(\mathbf{v})\overset{=}{*}0\right)\cdot\left(\overline{z}_j^k\overset{\neq}{*}\overline{z}_j^r\right)+\left(\overline{z}_j^k\overset{=}{*}\overline{z}_j^r\right)\right]}{\partial \mu_{A_{\mathrm{U}}^l}(\mathbf{v})}\right|_{l:\,\overline{z}_j^l=1}\right)\cdot \\ \cdot\sum\limits_{r=1}^{N}\mathop{T}\limits_{k=1}^{N}\left[\left(\mu_{A_{\mathrm{U}}^k}(\mathbf{v})\overset{=}{*}0\right)\cdot\left(\overline{z}_j^k\overset{\neq}{*}\overline{z}_j^r\right)+\left(\overline{z}_j^k\overset{=}{*}\overline{z}_j^r\right)\right] \\ -\sum\limits_{r=1}^{N}\overline{z}_j^r\cdot\mathop{T}\limits_{k=1}^{N}\left[\left(\mu_{A_{\mathrm{U}}^k}(\mathbf{v})\overset{=}{*}0\right)\cdot\left(\overline{z}_j^k\overset{\neq}{*}\overline{z}_j^r\right)+\left(\overline{z}_j^k\overset{=}{*}\overline{z}_j^r\right)\right]\cdot \\ \cdot\left(0+\sum\limits_{\substack{r=1\\ r:\,\overline{z}_j^r=-1}}^{N} \left.\frac{\partial \mathop{T}\limits_{k=1}^{N}\left[\left(\mu_{A_{\mathrm{U}}^k}(\mathbf{v})\overset{=}{*}0\right)\cdot\left(\overline{z}_j^k\overset{\neq}{*}\overline{z}_j^r\right)+\left(\overline{z}_j^k\overset{=}{*}\overline{z}_j^r\right)\right]}{\partial \mu_{A_{\mathrm{U}}^l}(\mathbf{v})}\right|_{l:\,\overline{z}_j^l=1}\right) \end{pmatrix}}{\left[\sum\limits_{r=1}^{N}\mathop{T}\limits_{k=1}^{N}\left[\left(\mu_{A_{\mathrm{U}}^k}(\mathbf{v})\overset{=}{*}0\right)\cdot\left(\overline{z}_j^k\overset{\neq}{*}\overline{z}_j^r\right)+\left(\overline{z}_j^k\overset{=}{*}\overline{z}_j^r\right)\right]\right]^2}, \tag{8.173}$$

$$\left.\frac{\partial \overline{z}_j}{\partial \mu_{A_{\mathrm{U}}^l}(\mathbf{v})}\right|_{l:\,\overline{z}_j^l=-1} = \frac{\begin{pmatrix} -\sum\limits_{\substack{r=1\\ r:\,\overline{z}_j^r=-1}}^{N} \left.\frac{\partial \mathop{T}\limits_{k=1}^{N}\left[\left(\mu_{A_{\mathrm{U}}^k}(\mathbf{v})\overset{=}{*}0\right)\cdot\left(\overline{z}_j^k\overset{\neq}{*}\overline{z}_j^r\right)+\left(\overline{z}_j^k\overset{=}{*}\overline{z}_j^r\right)\right]}{\partial \mu_{A_{\mathrm{U}}^l}(\mathbf{v})}\right|_{l:\,\overline{z}_j^l=1}\cdot \\ \cdot\sum\limits_{r=1}^{N}\mathop{T}\limits_{k=1}^{N}\left[\left(\mu_{A_{\mathrm{U}}^k}(\mathbf{v})\overset{=}{*}0\right)\cdot\left(\overline{z}_j^k\overset{\neq}{*}\overline{z}_j^r\right)+\left(\overline{z}_j^k\overset{=}{*}\overline{z}_j^r\right)\right] \\ -\sum\limits_{r=1}^{N}\overline{z}_j^r\cdot\mathop{T}\limits_{k=1}^{N}\left[\left(\mu_{A_{\mathrm{U}}^k}(\mathbf{v})\overset{=}{*}0\right)\cdot\left(\overline{z}_j^k\overset{\neq}{*}\overline{z}_j^r\right)+\left(\overline{z}_j^k\overset{=}{*}\overline{z}_j^r\right)\right]\cdot \\ \cdot\sum\limits_{\substack{r=1\\ r:\,\overline{z}_j^r=-1}}^{N} \left.\frac{\partial \mathop{T}\limits_{k=1}^{N}\left[\left(\mu_{A_{\mathrm{U}}^k}(\mathbf{v})\overset{=}{*}0\right)\cdot\left(\overline{z}_j^k\overset{\neq}{*}\overline{z}_j^r\right)+\left(\overline{z}_j^k\overset{=}{*}\overline{z}_j^r\right)\right]}{\partial \mu_{A_{\mathrm{U}}^l}(\mathbf{v})}\right|_{l:\,\overline{z}_j^l=1} \end{pmatrix}}{\left[\sum\limits_{r=1}^{N}\mathop{T}\limits_{k=1}^{N}\left[\left(\mu_{A_{\mathrm{U}}^k}(\mathbf{v})\overset{=}{*}0\right)\cdot\left(\overline{z}_j^k\overset{\neq}{*}\overline{z}_j^r\right)+\left(\overline{z}_j^k\overset{=}{*}\overline{z}_j^r\right)\right]\right]^2}, \tag{8.174}$$

$$\left.\frac{\partial \overline{z}_j}{\partial \mu_{A_{\mathrm{U}}^l}(\mathbf{v})}\right|_{l:\,\overline{z}_j^l=-1} = -\frac{\begin{pmatrix} \sum\limits_{r=1}^{N}\left(1+\overline{z}_j^r\right)\cdot\mathop{T}\limits_{k=1}^{N}\left[\left(\mu_{A_{\mathrm{U}}^k}(\mathbf{v})\overset{=}{*}0\right)\cdot\left(\overline{z}_j^k\overset{\neq}{*}\overline{z}_j^r\right)+\left(\overline{z}_j^k\overset{=}{*}\overline{z}_j^r\right)\right]\cdot \\ \cdot\sum\limits_{\substack{r=1\\ r:\,\overline{z}_j^r=-1}}^{N} \left.\frac{\partial \mathop{T}\limits_{k=1}^{N}\left[\left(\mu_{A_{\mathrm{U}}^k}(\mathbf{v})\overset{=}{*}0\right)\cdot\left(\overline{z}_j^k\overset{\neq}{*}\overline{z}_j^r\right)+\left(\overline{z}_j^k\overset{=}{*}\overline{z}_j^r\right)\right]}{\partial \mu_{A_{\mathrm{U}}^l}(\mathbf{v})}\right|_{l:\,\overline{z}_j^l=1} \end{pmatrix}}{\left[\sum\limits_{r=1}^{N}\mathop{T}\limits_{k=1}^{N}\left[\left(\mu_{A_{\mathrm{U}}^k}(\mathbf{v})\overset{=}{*}0\right)\cdot\left(\overline{z}_j^k\overset{\neq}{*}\overline{z}_j^r\right)+\left(\overline{z}_j^k\overset{=}{*}\overline{z}_j^r\right)\right]\right]^2}, \tag{8.175}$$

T–norm is nondecrasing, and operation $\left(\mu_{A_{\mathrm{U}}^k}(\mathbf{v})\overset{=}{*}0\right)$, which is a special case of negation, is nonincreasing, thus

$$\left.\frac{\partial \overline{z}_j}{\partial \mu_{A_{\mathrm{U}}^l}(\mathbf{v})}\right|_{l:\,\overline{z}_j^l=1} \geq 0, \tag{8.176}$$

Analyzing inequalities (8.155), (8.162), (8.169) and (8.176) it is easily seen that in formulas (8.145) and (8.146) fuzzy sets A_{L}^{r} and A_{U}^{r} should be chosen according to descriptions (8.147) and (8.148).

8.3.5 QL–DCOG Rough Fuzzy Classifier

It is easily seen that the lower and upper approximations of the membership of object x to class ω_j are given by

$$\underline{z}_j = \min_{\substack{A_{\mathrm{L}}^{k}=\left\{\underline{\widetilde{P}}A^k,\overline{\overline{\widetilde{P}}}A^k\right\}\\ k=1,\ldots,N}} \frac{\sum\limits_{r=1}^{N} \overline{z}_j^r \cdot \mathop{T}\limits_{k=1}^{N} S\left(N\left(\mu_{A_{\mathrm{L}}^{k}}(\mathbf{v})\right), \left(\mu_{A_{\mathrm{L}}^{k}}(\mathbf{v}) \cdot \left(\overline{z}_j^k \,\overline{\overline{*}}\, \overline{z}_j^r\right)\right)\right)}{\sum\limits_{r=1}^{N} \mathop{T}\limits_{k=1}^{N} S\left(N\left(\mu_{A_{\mathrm{L}}^{k}}(\mathbf{v})\right), \left(\mu_{A_{\mathrm{L}}^{k}}(\mathbf{v}) \cdot \left(\overline{z}_j^k \,\overline{\overline{*}}\, \overline{z}_j^r\right)\right)\right)}. \tag{8.177}$$

$$\overline{\overline{z}}_j = \max_{\substack{A_{\mathrm{U}}^{k}=\left\{\underline{\widetilde{P}}A^k,\overline{\overline{\widetilde{P}}}A^k\right\}\\ k=1,\ldots,N}} \frac{\sum\limits_{r=1}^{N} \overline{z}_j^r \cdot \mathop{T}\limits_{k=1}^{N} S\left(N\left(\mu_{A_{\mathrm{U}}^{k}}(\mathbf{v})\right), \left(\mu_{A_{\mathrm{U}}^{k}}(\mathbf{v}) \cdot \left(\overline{z}_j^k \,\overline{\overline{*}}\, \overline{z}_j^r\right)\right)\right)}{\sum\limits_{r=1}^{N} \mathop{T}\limits_{k=1}^{N} S\left(N\left(\mu_{A_{\mathrm{U}}^{k}}(\mathbf{v})\right), \left(\mu_{A_{\mathrm{U}}^{k}}(\mathbf{v}) \cdot \left(\overline{z}_j^k \,\overline{\overline{*}}\, \overline{z}_j^r\right)\right)\right)}. \tag{8.178}$$

In this case it is not possible to uniquely determine sets $\underline{\widetilde{P}}A^r$ and $\overline{\overline{\widetilde{P}}}A^r$ for subsystems describing lower and upper approximation of object membership to specific classes. Similarly as in [18], if there are N rules, the system would consist of 2^N subsystems described by (8.47), using all possible variations of approximations $\underline{\widetilde{P}}A^r$ and $\overline{\overline{\widetilde{P}}}A^r$.

8.3.6 MICOG Rough Fuzzy Classifier

The lower and upper approximations of the membership of object x to class ω_j are given by

$$\underline{z}_j = \frac{\sum\limits_{r=1}^{N} \overline{z}_j^r \cdot g^*\left(\mu_{A_{\mathrm{L}}^{r}}(\mathbf{v})\right)}{\sum\limits_{r=1}^{N} g^*\left(\mu_{A_{\mathrm{L}}^{r}}(\mathbf{v})\right)}, \tag{8.179}$$

$$\overline{\overline{z}}_j = \frac{\sum\limits_{r=1}^{N} \overline{z}_j^r \cdot g^*\left(\mu_{A_{\mathrm{U}}^{r}}(\mathbf{v})\right)}{\sum\limits_{r=1}^{N} g^*\left(\mu_{A_{\mathrm{U}}^{r}}(\mathbf{v})\right)}, \tag{8.180}$$

Theorem 8.5. *Fuzzy sets A_{L}^{r} and A_{U}^{r} in descriptions* (8.179) *and* (8.180) *should be chosen as follows*

$$A_{\mathrm{L}}^{r} = \begin{cases} \underline{\widetilde{P}}A^r & \text{if } \overline{\mathrm{z}}_{\mathrm{j}}^{\mathrm{r}} = 1 \\ \overline{\overline{\widetilde{P}}}A^r & \text{if } \overline{\mathrm{z}}_{\mathrm{j}}^{\mathrm{r}} = -1. \end{cases} \tag{8.181}$$

$$A_{\mathrm{U}}^{r} = \begin{cases} \overline{\overline{P}}A^{r} & \text{if } \overline{z}_{\mathrm{j}}^{\mathrm{r}} = 1 \\ \underline{\widetilde{P}}A^{r} & \text{if } \overline{z}_{\mathrm{j}}^{\mathrm{r}} = -1. \end{cases} \tag{8.182}$$

Proof. This theorem can be proved in the same way as that in [15].

8.4 Ensembles

We will now combine rough-fuzzy classifiers, described in the previous chapter, by means of the AdaBoost algorithm. The pair of outputs of the each system, i.e. $\left(\underline{\overline{z}_j}, \overline{\overline{z}_j}\right)$ is treated as the interval. When values of all input features are known, there is only one output value $\overline{z}_j = \underline{\overline{z}_j} = \overline{\overline{z}_j}$. When some values are missing, the output value, which is achieved if all features are known, belongs to obtained interval, i.e.

$$\underline{\overline{z}_j} < \overline{z}_j < \overline{\overline{z}_j} \tag{8.183}$$

To compute the overall output of the ensemble of classifiers trained by AdaBoost algorithm the following formula is used $f(\mathbf{x}) = \sum_{t=1}^{T} c_t h_t(\mathbf{x})$, where $c_t = \frac{\alpha_t}{\sum_{t=1}^{T} \alpha_t}$ is classifier importance for a given training set, $h_t(\mathbf{x})$ is the response of the hypothesis t on the basis of feature vector $\mathbf{x} = [x_1, ..., x_n]$. The coefficient c_t value is computed on the basis of the classifier error and can be interpreted as the measure of classification accuracy of the given classifier. As we see, the AdaBoost algorithm is a meta-learning algorithm and does not determine the way of learning for classifiers in the ensemble.

Let us fix two numbers (thresholds) z_{IN} and z_{OUT} such that $1 > z_{\mathrm{IN}} \geq q z_{\mathrm{OUT}} > 0$. Then the decision of a single classifier can be defined as follows

$$h_t(x) = \begin{cases} 1 & \text{if } \underline{\overline{z}_j} \geq z_{\mathrm{IN}} \text{ and } \overline{\overline{z}_j} > z_{\mathrm{IN}} \\ -1 & \text{if } \underline{\overline{z}_j} < z_{\mathrm{OUT}} \text{ and } \overline{\overline{z}_j} \leq z_{\mathrm{OUT}} \\ \frac{1}{2} & \text{if } z_{\mathrm{IN}} > \underline{\overline{z}_j} \geq z_{\mathrm{OUT}} \text{ andi } \overline{\overline{z}_j} > z_{\mathrm{IN}} \\ -\frac{1}{2} & \text{if } \underline{\overline{z}_j} < z_{\mathrm{OUT}} \text{ and } z_{\mathrm{OUT}} < \overline{\overline{z}_j} \leq z_{\mathrm{IN}} \\ 0 & \text{in other cases.} \end{cases} \tag{8.184}$$

or in a simpler version

$$h_t(x) = \begin{cases} 1 & \text{if } \underline{\overline{z}_j} \geq z_{\mathrm{IN}} \text{and } \overline{\overline{z}_j} > z_{\mathrm{IN}} \\ -1 & \text{if } \underline{\overline{z}_j} < z_{\mathrm{OUT}} \text{ and } \overline{\overline{z}_j} \leq z_{\mathrm{OUT}} \\ 0 & \text{in other cases.} \end{cases} \tag{8.185}$$

The zero output value of the t–th classifier is interpreted as a refusal to classify because of too small number of features. We use rules from the ensemble of neuro-fuzzy systems in the rough-neuro-fuzzy classifier. If we want to consider, in the overall ensemble response, only the classifiers which do not refuse to provide the answer, we redefine (3.9), taking into account (3.10), to get the following form:

$$f(x) = \begin{cases} \dfrac{\sum\limits_{\substack{t=1 \\ :h_t(x)\neq 0}}^{T} \alpha_t h_t(x)}{\sum\limits_{\substack{t=1 \\ :h_t(x)\neq 0}}^{T} \alpha_t} & \text{if } \sum\limits_{\substack{t=1 \\ :h_t(x)\neq 0}}^{T} \alpha_t > 0 \\ 0 & \text{if } \sum\limits_{\substack{t=1 \\ :h_t(x)\neq 0}}^{T} \alpha_t = 0 \end{cases} \tag{8.186}$$

$$H(x) = sgn\left(f(x)\right) \tag{8.187}$$

The output value can be interpreted as follows

- $H(x) = 1$ — object belongs to the class,
- $H(x) = 0$ — the ensemble does not know,
- $H(x) = -1$ — object does not belong to the class,

if we assume that $z_{\text{IN}} = z_{\text{OUT}} = 0$.

8.5 Experimental Results

In this section we test the algorithms proposed in this chapter using two well known benchmarks from [6]. We learned the neuro-fuzzy systems using the backpropagation algorithm combined with the AdaBoost (Section 3.2), then we used obtained fuzzy rules in the rough-neuro-fuzzy ensembles. In the experiments we have tested the cases when the part of feature values was unavailable. More specifically, we have examined all possible combinations of available and not available values. All

Table 8.2 Ensemble performance on learning/testing on the WBCD benchmark for Mamdani rough–neuro-fuzzy ensemble (Section 8.3.1) when the number of features is decreasing

number of available features	T–CA rough neuro-fuzzy ensemble on WBCD correct class. [%]	no class. [%]	incorrect class. [%]
9	96.79 / 97.45	0.00 / 0.00	3.21 / 2.55
8	91.48 / 92.17	7.02 / 6.08	1.5 / 1.74
7	46.16 / 45.92	52.99 / 52.91	0.85 / 1.17
6	27.29 / 26.73	72.13 / 72.52	0.59 / 0.75
5	20.89 / 20.47	78.78 / 79.14	0.33 / 0.39
4	12.98 / 12.88	86.91 / 87.00	0.10 / 0.12
3	5.35 / 5.52	94.64 / 94.47	0.01 / 0.01
2	0.87 / 0.88	99.13 / 99.12	0.00 / 0.00
1	0.00 / 0.00	100.00 / 100.00	0.00 / 0.00
0	0.00 / 0.00	100.00 / 100.00	0.00 / 0.00

Table 8.3 Ensemble performance on learning/testing on the PID benchmark for Mamdani rough–neuro-fuzzy ensemble (Section 8.3.1) when the number of features is decreasing

number of	T-CA rough neuro-fuzzy ensemble on PID		
available	correct class.	no class.	incorrect class.
features	[%]	[%]	[%]
8	77.08 / 76.91	0.00 / 0	22.92 / 23.09
7	51.37 / 52.08	35.74 / 34.61	12.89 / 13.30
6	22.34 / 20.32	74.65 / 76.13	3.01 / 3.55
5	5.14 / 3.87	94.46 / 95.60	0.40 / 0.52
4	0.85 / 0.36	99.05 / 99.51	0.10 / 0.12
3	0.23 / 0.07	99.74 / 99.89	0.03 / 0.03
2	0.00 / 0.00	100.00 / 100.00	0.00 / 0.00
1	0.00 / 0.00	100.00 / 100.00	0.00 / 0.00
0	0.00 / 0.00	100.00 / 100.00	0.00 / 0.00

Table 8.4 Ensemble performance on learning/testing on the WBCD benchmark for QL-DCOG rough–neuro-fuzzy ensemble (Section 8.3.5) when the number of features is decreasing

number of	QL-DCOG rough neuro-fuzzy ensemble on WBCD		
available	correct class.	no class.	incorrect class.
features	[%]	[%]	[%]
9	97.81 / 97.44	0.00 / 0.00	2.19 / 2.56
8	80.49 / 79.40	17.96 / 19.38	1.54 / 1.22
7	37.81 / 38.17	61.23 / 61.13	0.95 / 0.70
6	21.69 / 22	78.02 / 77.68	0.30 / 0.32
5	12.71 / 12.43	87.26 / 87.50	0.03 / 0.06
4	4.10 / 3.86	95.90 / 96.13	0.00 / 0.00
3	0.27 / 0.33	99.73 / 99.67	0.00 / 0.00
2	0.00 / 0.00	100.00 / 100.00	0.00 / 0.00
1	0.00 / 0.00	100.00 / 100.00	0.00 / 0.00
0	0.00 / 0.00	100.00 / 100.00	0.00 / 0.00

tables with simulation results are obtained for every combination of missing features, namely for all available features, for one missing feature, for two missing features and so on. For example, Table 8.2 has ten rows for each possible combination of missing features. When all features are available (first row) and classification accuracy is equal to 96.79% and 97.45% for learning and testing sequences, respectively. The second row contains average of nine results, each when one feature was removed (eight features were available in each run). In this case, corresponding accuracy for testing data set was 92.17%, 6.08% of objects were not classified and 1.74% of objects were classified incorrectly. When the number of not available features increases, incorrect classification ratio is not increased. Instead, no classification ratio increases, what gives us the indication that the number of features needed

Table 8.5 Ensemble performance on learning/testing on the WBCD benchmark for MICOG Kleene-Dienes rough–neuro-fuzzy ensemble (Section 8.3.6) when the number of features is decreasing

number of	MICOG Kleene-Dienes rough neuro-fuzzy ensemble WBCD		
available features	correct class. [%]	no class. [%]	incorrect class. [%]
9	97.45 / 97.16	0.00 / 0.00	2.55 / 2.84
8	68.21 / 67.24	30.01 / 30.89	1.78 / 1.87
7	35.08 / 34.58	63.48 / 64.06	1.44 / 1.35
6	30.12 / 30.00	68.69 / 69.01	1.19 / 0.99
5	26.11 / 26.35	73.06 / 72.9	0.83 / 0.75
4	20.43 / 20.66	79.11 / 78.83	0.46 / 0.51
3	13.24 / 13.57	86.53 / 86.13	0.23 / 0.30
2	5.96 / 6.18	93.95 / 93.69	0.09 / 0.13
1	0.00 / 0.00	100.00 / 100.00	0.00 / 0.00
0	0.00 / 0.00	100.00 / 100.00	0.00 / 0.00

Table 8.6 MICOG Łukasiewicz ensemble performance on learning/testing on the WBCD benchmark for Mamdani rough–neuro-fuzzy ensemble (Section 8.3.6) when the number of features is decreasing

number of	MICOGŁukasiewicz rough neuro-fuzzy ensemble WBCD		
available features	correct class. [%]	no class. [%]	incorrect class. [%]
9	96.72 / 96.79	0.00 / 0.00	3.28 / 3.21
8	79.40 / 79.57	18.98 / 18.80	1.62 / 1.63
7	40.91 / 41.40	58.04 / 57.60	1.05 / 1.00
6	25.97 / 26.74	73.35 / 72.58	0.67 / 0.68
5	20.28 / 21.07	79.37 / 78.56	0.35 / 0.37
4	13.90 / 14.23	85.92 / 85.64	0.17 / 0.13
3	7.16 / 7.09	92.79 / 92.89	0.06 / 0.02
2	1.87 / 1.80	98.12 / 98.20	0.01 / 0.00
1	0.00 / 0.00	100.00 / 100.00	0.00 / 0.00
0	0.00 / 0.00	100.00 / 100.00	0.00 / 0.00

to perform classification with the desired accuracy should be increased. It should be noted that the results are given for learning/testing data set.

We used five neuro–fuzzy systems each with two fuzzy rules to classify the Wisconsin Breast Cancer Database problem (Section 1.2.4). Performance results of rough–neuro-fuzzy ensembles for changing number of features are presented in Tables 8.2, 8.4, 8.5, 8.6, 8.7. We also used five neuro–fuzzy systems each with two fuzzy rules to classify the Pima Indians diabetes problem (Section 1.2.3). Obtained rules are used in the rough–neuro-fuzzy ensemble. The classification results for changing number of available features are presented in Table 8.3.

Table 8.7 T-DCOG ensemble performance on learning/testing on the WBCD benchmark for Mamdani rough–neuro-fuzzy ensemble (Section 8.3.2) when the number of features is decreasing

number of available features	T-DCOG rough neuro-fuzzy ensemble on WBCD correct class. [%]	no class. [%]	incorrect class. [%]
9	97,45 / 97,34	0 / 0	2,55 / 2,66
8	97 / 96,6	0 / 0	3 / 3,4
7	95,55 / 95,5	0,77 / 0,68	3,68 / 3,82
6	84,72 / 85,03	13,11 / 12,89	2,17 / 2,08
5	47,03 / 47,23	52,26 / 52,14	0,71 / 0,63
4	17,61 / 17,92	82,12 / 81,84	0,27 / 0,24
3	7,82 / 7,82	92,12 / 92,13	0,06 / 0,04
2	1,7 / 1,72	98,3 / 98,28	0 / 0
1	0 / 0	100 / 100	0 / 0
0	0 / 0	100 / 100	0 / 0

8.6 Summary and Discussion

Neural networks are able to perfectly fit to data and fuzzy logic systems, presented in previous chapters, use interpretable knowledge. These methods cannot handle data with missing or unknown features what can be achieved easily using rough set theory. Thus, in this chapter we proposed a new class of modular systems for classification in the case of missing feature values. The rough set theory was incorporated to ensembles of neuro–fuzzy systems to perform classification even if some features are missing. The ensembles were created by the AdaBoost metalearning algorithm which was combined with the backpropagation algorithm to train neuro-fuzzy systems. Fuzzy rules from the trained ensemble of neuro-fuzzy systems were used to build the knowledge base for a rough-neuro-fuzzy modular classifier. The accuracy of the rough-neuro-fuzzy modular system is high and the system is able to classify data with missing features, and even provide information that the number of available features is too small and proper classification is impossible. The fundamental results of this chapter were partly presented in [8, 29].

Experiments showed very clearly the good accuracy of the system and the ability to work when the number of available features decreases. In the numerical experiments, fuzzy set membership functions are set in advance by the fuzzy *c*-means algorithm. Then, all system parameters are determined by the backpropagation algorithm. Boosting weights influence the backpropagation learning by inhibiting fuzzy system parameter modification for a given sample.

References

1. Broersen, P.M.T., de Waele, S., Bos, R.: Application of autoregressive spectral analysis to missing data problems. IEEE Transactions on Instrumentation and Measurement 53, 981–986 (2004)

2. Chan, L.S., Dun, O.J.: Alternative approaches to missing values in discriminant analysis. Journal of the American Statistical Association 71, 842–844 (1976)
3. Cooke, M., Green, P., Josifovski, L., Vizinho, A.: Robust automatic speech recognition with missing and unreliable acoustic data. Speech Communication 34(3), 267–285 (2001)
4. Dixon, J.K.: Pattern recognition with partly missing data. IEEE Transactions on Systems, Man, and Cybernetics 9(10), 617–621 (1979)
5. Duda, R.O., Hart, P.E., Stork, D.G.: Pattern Classification. Wiley-Interscience Publication (2000)
6. Frank, A., Asuncion, A.: UCI machine learning repository (2010), `http://archive.ics.uci.edu/ml`
7. Jang, R.J.S., Sun, C.T., Mizutani, E.: Neuro-Fuzzy and Soft Computing, A Computational Approach to Learning and Machine Intelligence. Prentice-Hall, Upper Saddle River (1997)
8. Korytkowski, M., Nowicki, R., Scherer, R.: Neuro-fuzzy Rough Classifier Ensemble. In: Alippi, C., Polycarpou, M., Panayiotou, C., Ellinas, G. (eds.) ICANN 2009. LNCS, vol. 5768, pp. 817–823. Springer, Heidelberg (2009)
9. Korytkowski, M., Scherer, R., Rutkowski, L.: On combining backpropagation with boosting. In: 2006 International Joint Conference on Neural Networks, IEEE World Congress on Computational Intelligence, Vancouver, BC, Canada, pp. 1274–1277 (2006), doi:10.1109/IJCNN.2006.246838
10. Kuncheva, L.: Fuzzy Classifier Design. STUDFUZZ. Physica-Verlag, Heidelberg (2000)
11. Kuncheva, L.: Combining Pattern Classifiers. STUDFUZZ. John Wiley & Sons (2004)
12. Mas, M., Monserrat, M., Torrens, J.: Two types of implications derived from uninorms. Fuzzy Sets and Systems 158, 2612–2626 (2007)
13. Morin, R.L., Raeside, D.E.: A reappraisal of distance-weighted k-nearest neighbor classification for pattern recognition with missing data. IEEE Transactions on Systems, Man, and Cybernetics 11, 241–243 (1981)
14. Nauck, D.: Foundations of Neuro-Fuzzy Systems. John Wiley, Chichester (1997)
15. Nowicki, R.: On combining neuro-fuzzy architectures with the rough set theory to solve classification problems with incomplete data. IEEE Trans. Knowl. Data Eng. 20(9), 1239–1253 (2008), doi:10.1109/TKDE.2008.64
16. Nowicki, R.: Rough-neuro-fuzzy structures for classification with missing data. IEEE Trans. Syst., Man, Cybern. B 39, 1334–1347 (2009)
17. Nowicki, R.: On classification with missing data using rough-neuro-fuzzy systems. International Journal of Applied Mathematics and Computer Science 20(1), 55–67 (2010), doi:10.2478/v10006-010-0004-8
18. Nowicki, R.K.: Fuzzy decision systems for tasks with limited knowledge. Academic Publishing House EXIT (2009) (in polish)
19. Pawlak, M.: Kernel classification rules from missing data. IEEE Transactions on Information Theory 39, 979–988 (1993)
20. Pawlak, Z.: Rough sets. International Journal of Information and Computer Science 11(341), 341–356 (1982)
21. Pawlak, Z.: Rough Sets: Theoretical Aspects of Reasoning About Data. Kluwer, Dordrecht (1991)
22. Pawlak, Z.: Rough sets, decision algorithms and bayes' theorem. European Journal of Operational Research 136, 181–189 (2002)
23. Renz, C., Rajapakse, J.C., Razvi, K., Liang, S.K.C.: Ovarian cancer classification with missing data. In: Proceedings of the 9th International Conference on Neural Information — ICONIP 2002, vol. 2, pp. 809–813 (2002)

24. Rutkowska, D., Nowicki, R.: Implication - based neuro-fuzzy architectures. International Journal of Applied Mathematics and Computer Science 10(4), 675–701 (2000)
25. Rutkowski, L.: Flexible Neuro-Fuzzy Systems. Kluwer Academic Publishers (2004)
26. Rutkowski, L.: Computational Intelligence Methods and Techniques. Springer, Heidelberg (2008)
27. Rutkowski, L., Cpałka, K.: Flexible neuro-fuzzy systems. IEEE Trans. Neural Networks 39(3), 554–574 (2003)
28. Rutkowski, L., Cpałka, K.: Designing and learning of adjustable quasi-triangular norms with applications to neuro-fuzzy systems. IEEE Trans. Fuzzy Systems 13(1), 140–151 (2005)
29. Scherer, R., Korytkowski, M., Nowicki, R., Rutkowski, L.: Modular Rough Neuro-fuzzy Systems for Classification. In: Wyrzykowski, R., Dongarra, J., Karczewski, K., Wasniewski, J. (eds.) PPAM 2007. LNCS, vol. 4967, pp. 540–548. Springer, Heidelberg (2008), `http://www.springerlink.com/content/372n3787p08x817m/`
30. Tanaka, M., Kotokawa, Y., Tanino, T.: Patern classification by stochastic neural network with missing data. In: Proceedings of IEEE International Conference on System, Man and Cybernetics, vol. 1, pp. 690–695 (1996)
31. Wang, L.X.: Adaptive Fuzzy Systems and Control. PTR Prentice-Hall, Englewood Cliffs (1994)

Chapter 9
Concluding Remarks and Challenges for Future Research

The results presented in this book lead us toward improving certain areas of classification. In the book two difficult problems are solved. It was shown how to join fuzzy rules from all subsystems creating an ensemble and how to design an ensemble of fuzzy subsystems in the case of incomplete data. In particular the book contributed with a new method of ensemble backpropagation learning that takes into account boosting weights, modification of the fuzzy c-means clustering algorithm for ensembles, novel design of the Mamdani, Takagi Sugeno, relational and logical fuzzy systems constituting the ensemble, resulting in normalization of individual rule bases during learning and a family of various rough-neuro-fuzzy ensembles. Novel and original methods of learning suited for ensembles and new methods of creating ensembles are presented in Sections 3.4 and 3.5 and chapters 4–8. Chapter 3 presented some new ensemble techniques used in the book. Chapter 4 contained description and novel algorithms for relational neuro-fuzzy systems and methods to join them into larger ensembles. Chapters 5–7 presented ensembles consisted of, respectively, the Mamdani, logical and the Takagi-Sugeno fuzzy systems. Rough-neuro-fuzzy systems, used to build boosting ensembles, were shown in Chapter 8. The proposed algorithms can be applied in many domains such as medicine, economics, fraud or fault detection, etc. and more generally in decision making or pattern recognition.

Now we would like to identify some future research directions. The presented methods can be modified to use some extensions of fuzzy sets, e.g. type-2 fuzzy sets [7], interval valued fuzzy sets [6], intutionistic sets [1] or twofold fuzzy sets [3]. Presented approaches for incomplete data (Chapter 8) can be further modified using the rough set theory extensions such as variable precision rough sets [8] or dominance based rough sets [5]. Finally, we can use more sophisticated boosting variations, e.g. linear programming boosting [2] or BrownBoost [4].

R. Scherer: Multiple Fuzzy Classification Systems, STUDFUZZ 288, pp. 129–130.
springerlink.com

References

1. Atanassov, K.: Intuitionistic fuzzy sets. Theory and applications. Physica-Verlag, Heidelberg (1999)
2. Demiriz, A., Bennett, K.P., Shawe-Taylor, J.: Linear programming boosting via column generation. Mach. Learn. 46(1-3), 225–254 (2002), http://dx.doi.org/10.1023/A:1012470815092, doi:10.1023/A:1012470815092
3. Dubois, D., Prade, H.: Twofold fuzzy sets and rough sets – some issues in knowledge representation. Fuzzy Sets and Systems 23(1), 3–18 (1987)
4. Freund, Y.: An adaptive version of the boost by majority algorithm. In: Proceedings of the Twelfth Annual Conference on Computational Learning Theory, pp. 102–113 (2000)
5. Greco, S., Matarazzo, B., Slowinski, R.: Rough sets theory for multicriteria decision analysis. European Journal of Operational Research 129, 1–47 (2001)
6. Zadeh, L.: Outline of a new approach to analysis of complex systems and decision processes. IEEE Trans. Syst Man Cybern 3, 28–44 (1973)
7. Zadeh, L.A.: The concept of a linguistic variable and its application to approximate reasoning. Inf. Sci. 8(3), 199–249 (1975)
8. Ziarko, W.: Variable precision rough set model. Journal of Computer and System Sciences 46(1), 39–59 (1993)

Index